A Business History of the Swatch Group

Also by Pierre-Yves Donzé

ORGANIZING GLOBAL TECHNOLOGY FLOWS: Institutions, Actors, and Processes (*edited with Shigehiro Nishimura*)

FROM FAMILY BUSINESS TO GLOBAL BRAND: Longines

HISTORY OF THE SWISS WATCH INDUSTRY FROM JACQUES DAVID TO NICOLAS HAYEK

LES PATRONS HORLOGERS DE LA CHAUX-DE-FONDS (1840–1920): Dynamique sociale d'une élite industrielle

L'OMBRE DE CÉSAR: Les chirurgiens et la construction du système hospitalier vaudois (1840–1960)

A Business History of the Swatch Group

The Rebirth of Swiss Watchmaking and the Globalization of the Luxury Industry

Pierre-Yves Donzé
Graduate School of Economics, Kyoto University, Japan

palgrave
macmillan

First published 2014 by
PALGRAVE MACMILLAN

Palgrave Macmillan in the UK is an imprint of Macmillan Publishers Limited, registered in England, company number 785998, of Houndmills, Basingstoke, Hampshire RG21 6XS.

Palgrave Macmillan in the US is a division of St Martin's Press LLC, 175 Fifth Avenue, New York, NY 10010.

Palgrave Macmillan is the global academic imprint of the above companies and has companies and representatives throughout the world.

Palgrave® and Macmillan® are registered trademarks in the United States, the United Kingdom, Europe and other countries

ISBN: 978–1–137–38906–0

This book is printed on paper suitable for recycling and made from fully managed and sustained forest sources. Logging, pulping and manufacturing processes are expected to conform to the environmental regulations of the country of origin.

A catalogue record for this book is available from the British Library.

A catalog record for this book is available from the Library of Congress.

Transferred to Digital Printing in 2014

Contents

List of Figures

List of Tables

Preface and Acknowledgements

For more than two decades, the luxury industry has posted one of the fastest growth rates among all economic sectors – a trend which most financial analysts expect to continue. According to the consulting firm Bain & Company, for example, global luxury sales exploded in less than two decades, rising from 77 billion euros in 1995 to 212 billion in 2012.[1] However, the companies producing luxury goods – champagne, fashion articles, perfume, jewellery, and watches – have undergone a radical change since the 1980s that is not always easy to define, as a result of a communication strategy designed precisely to reinforce the timeless nature of the industry and its products. History, tradition, and legends are the essential resources for marketing and added value in the luxury industry, so much so that it is often difficult to go beyond conventional brand language to analyse the development of these firms and this industry.

This book takes as an example watchmaking, an industrial sector whose recent dynamic performance is relatively unexplored. Of course, the Swiss and global watchmaking industry, from the 18th century up to the sea changes of the 1960s and 1970s, has been the subject of many works, which provide a clear explanation of the key points of the history of this economic sector over the long term.[2] Yet recent developments in this industry, from the crisis of 1975–1985 to the present, are not well known, and paradoxically have not given rise to many in-depth studies. There are two – perhaps surprising – reasons for this lack of interest. First, it is a reflection of the specific characteristics of Swiss academia: historians rarely extend their research up to the present moment, and economists display a certain lack of interest in applied economic approaches. Second, it mirrors the attitude of the media, as few watchmaking journalists compile thorough analyses of developments in the industry, limiting themselves to passing on the viewpoint of the communication departments of the major watchmaking groups. As a result, the history of the Swiss watchmaking industry over the past three decades is often confined to images which have become virtually legendary: a revolutionary product (the Swatch), a brilliant entrepreneur (Nicolas Hayek), and a time-honoured culture particular to the Swiss Jura Arc (mechanical watchmaking).

We must look beyond these images to determine how Swiss watchmakers have staged a comeback on world markets since the late 1980s. There are several viable approaches to tackling this topic. This book chooses to deal with the Swatch Group and offers an applied business history analysis inspired by the work of Takeo Kikkawa. In his publications covering the Japanese petroleum industry, Kikkawa showed how a historical perspective was useful for understanding firms' current ability (or inability) to compete.[3] An advocate of a historical approach extending up to the present, he defined his perspective as follows: "The study of applied business history is about identifying the dynamism of industrial and corporate growth through an examination of business history; then, based on the findings, such studies explore solutions to the contemporary problems encountered by relevant industries and companies."[4] In the case of this analysis of the Swiss watchmaking industry, even if the objective is purely cognitive and thus significantly different, on the grounds set out above such an approach firmly rooted in the present appears both relevant and necessary.

There are two main reasons for choosing the Swatch Group as the subject of this book. First, it is the world's leading watchmaking group. In Switzerland, it has employed approximately 25–30 per cent of all those working in the industry over the past 30 years (26.6 per cent of all Swiss watchmaking jobs in 1990; 28.9 per cent in 2000). Accordingly, the firm has considerable influence within the industry as a whole.

Second, the Swatch Group has a virtual monopoly – which is much more a relic of the past than a deliberate choice, as we will see – on the production of movement blanks (*ébauches*) and movement components through its subsidiary ETA SA, whose headquarters are in Granges. Thus, for historical rather than legal reasons (because Swiss legislation on monopolies has never been very restrictive from an international perspective), the Swatch Group supplies not only its own watchmaking factories but also those of the main rival groups, including some prestigious *manufactures* (watchmaking workshops). As a result of its control over movement-related technologies, the Swatch Group is a leader in this industry.

Casting an analytical eye over the modern watchmaking industry is not an easy task. It poses a number of methodological problems, the most difficult of which is gaining access to primary sources. Most companies, including the Swatch Group, do not grant access to their archives and publish only a limited amount of data in their annual reports. Even though they are meant for investors, these documents turned into communication tools during the 2000s and now contain less and less quantitative information that might be of use in understanding how a company works. By way of example, in 2005 the Swatch Group stopped publishing information on its sales volume (number of watches and watch movements) and the share of its employees in Asia. However, these limitations do not preclude an analysis of the modern Swiss watchmaking industry. The main documents used for this study are the Swatch Group's annual reports and official Swiss foreign trade statistics. To avoid making the text excessively lengthy and awkward, references are generally not given for information taken from these sources.

This book is an expanded and revised version in English of a book originally published in French in 2012 by Éditions Alphil-Presses Universitaires Suisses under the title *Histoire du Swatch Group*. Draft versions of the manuscript which formed the basis of this book were prepared as working papers at the University of Osaka (in English) and at the University of Neuchâtel (in French), as well as in the form of an article published in the *Journal of Strategic Management Education* under the title "The Comeback of the Swiss Watch Industry on the World Market: A Business History of the Swatch Group (1983–2010)" (vol. 8, no. 2, 2012, pp. 1–32). Finally, the bulk of Chapter 2 is taken from an essay entitled "Global Competition and Technological Innovation: A New Interpretation of the Watch Crisis, 1970s–1980s," published in 2012 as part of a collection edited by Thomas David, Jon Mathieu, Janick Marina Schaufelbuehl, and Tobias Straumann, *Crises – Causes, interprétations et consequences* (Zurich: Chronos, pp. 275–289). Previous drafts of this book benefited from the close reading, criticism, and comments of many people, to whom I am very grateful. My special thanks go to Nicolas Babey, Jean-Claude Biver, Bastien Buss, Olivier Crevoisier, Nicolas Hanssens, Laurence Marti, Christelle Roks, and Minoru Sawai. I also wish to express my gratitude to the Federation of the Swiss Watch Industry (FHS) for providing me with a great many statistics on watch exports. My heartfelt thanks go to my friend Alain Cortat, the publisher of the French version, for providing constructive criticism and advice. Finally, I want to thank David Neal for his kind support in translating this book and making it available to a worldwide audience. However, as author, I remain solely responsible for the views expressed and any errors.

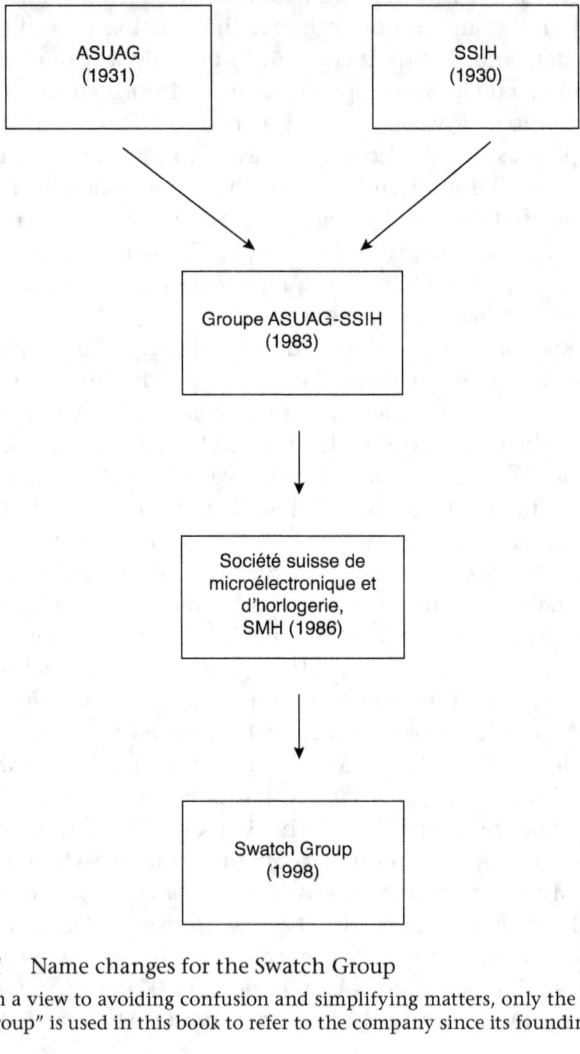

Figure 0.1 Name changes for the Swatch Group

Note: With a view to avoiding confusion and simplifying matters, only the name "the Swatch Group" is used in this book to refer to the company since its founding in 1983.

Notes

1. Bain & Company, *2012 Luxury Goods*. Bain's analysis covers only spending on luxury apparel, accessories, leather goods, shoes, jewellery, watches, perfume, and cosmetics. It excludes luxury services, hotels, restaurants, wines and spirits, luxury cars, and yachts/private jets.
2. See, for example, Landes, *Revolution in Time*; Glasmeier, *Manufacturing Time*; and Donzé, *History of the Swiss Watch Industry*.
3. Kikkawa, *Nihon sekiyu sangyo*.
4. Kikkawa, "International competitiveness", forthcoming.

1
Introduction

People were virtually writing the Swiss watchmaking industry off in the early 1980s, as it had not managed to contain the worldwide expansion of it's Japanese competitors. Powered by the mass production of first high-quality mechanical watches, then quartz watches, the Japanese watchmaking industry launched a growth policy in the second half of the 1960s in a bid to challenge Swiss domination of world markets.[1]

As a result of the extraordinary development of Japanese watchmaking companies in the 1970s, whose production value shot up from USD350 million in 1970 to 2 billion in 1980, Japan managed to overtake Switzerland in 1981–1985 (see Figure 1.1). During this five-year period, the average annual value of Japanese watch production was USD1.96 billion, of which 1.74 billion was for exports, compared with USD1.69 billion for Swiss watch exports (the production value of Swiss companies is not known, but is close to the value of exports, given the small size of the Swiss market, in which sales accounts for some 5 per cent of production). As a result, Japanese competition pushed the watchmaking industry in Switzerland into a crisis situation, marked by a decline and then stagnation in exports as well as a drop in employment. From the end of the 1980s, however, the dynamics were completely different: Swiss exports were booming (USD4.9 billion in 1990, 6.1 billion in 2000, and 16.2 billion in 2010), while Japan's watch production stagnated, then fell sharply (USD2.8 billion in 1990, 1.5 billion in 2000, and 0.9 billion in 2010).

Japanese researchers working in the management field have focused on the stagnation and decline of the Japanese watchmaking

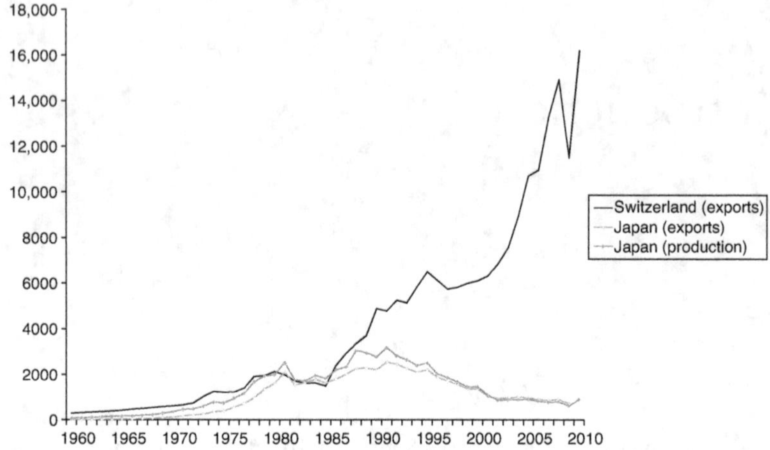

Figure 1.1 Production and exports of the Swiss and Japanese watchmaking industries, in USD millions, 1960–2010

Note: No figures are available for Swiss production.

Source: *Swiss annual foreign trade statistics*. Bern: Swiss Federal Customs Administration, 1960–2010; *Kikai tokei nenpo*. Tokyo: MITI, 1960–2010; *Nihon gaikoku boeki tokei*. Tokyo: Ministry of Finance, 1960–2010.

industry. Their work consistently focus on technology as the key factor to understand the dynamics of this industry. Indeed, the Japanese literature of the 1990s and 2000s emphasizes the diversification of products and the transformation of their architecture (the so-called modularization process) as strategies adopted in the hope of regaining lost competitiveness.[2] In so doing, they are in total agreement with the actors of the Japanese watchmaking industry, who remain narrowly obsessed by technological innovation. Designing new products and cutting production costs remain the two core strategies which Japanese watchmaking companies have been following, unsuccessfully so far, in their bid to reconquer world markets. On the one hand, they have successively launched genuinely innovative products, such as solar watches (1995), hybrid quartz-mechanism spring-drive watches (1999), and radio-controlled watches (2005), thereby asserting their innovativeness, which remains a key element of their communication strategy. On the other hand, they have relocated a great deal of production outside Japan, mainly to China, in an attempt to reduce manufacturing costs. As a result, the proportion of watches produced abroad by Japanese watchmaking companies has risen sharply, shooting up from 17.8 per cent in 1995 to 24.2 per cent in 2000 and 45.8 per cent in 2010.[3] In the final analysis, the aim of this twofold strategy is to bring innovative, inexpensive, and high-quality products to market, in the hope of once again competing with Swiss watchmakers. However, these measures have not helped Japanese watchmaking firms grow or become more competitive.

Understanding the rebirth of the Swiss watchmaking industry

Whereas Japanese researchers and industrialists attach great importance to technological innovation as a driver of industrial growth and corporate competiveness, Western management researchers and historians explain the extraordinary comeback of Swiss watchmaking on world markets in rather different terms. There are two main approaches. The first, which predominates, may be called the "cultural school" approach. In terms of explanations, it focuses on the industrial district. From this perspective, recourse to a few key elements of the Marshallian district (such as industrial climate and relations of competition–complementarity between companies) makes it possible to interpret the rebirth of the Swiss watchmaking industry as the outcome of a technical culture inscribed in the territory, which has made it possible to reposition the industry at the high end of the market and to manufacture quality products that rely on traditional know-how.[4] For example, Hélène Pasquier argues that "watchmaking firms positioned themselves as the guarantors of the age-old regional tradition."[5] This type of reasoning, which is quite widespread and is based on a "romantic description of the district,"[6] as the French sociologist Pierre-Paul Zalio puts it, is very close to the narrative produced by the watchmaking industry itself. Moreover, it attaches considerable importance to the Swatch's role in overcoming the crisis. This innovative product supposedly not only helped the Swiss watchmaking industry prevail against Japan but also enabled it, through the profits earned, to reposition itself to high-quality mechanical watchmaking, through a process which is never clearly exposed in this literature. However, this approach poses a problem, first because it proposes a model which is based on a debatable interpretation of the Marshallian district and an insufficient analysis of the sources, and second because it fails to situate the Swiss watchmaking industry within the global context.

The second approach explains the rebirth of the Swiss watchmaking industry as the consequence of a new marketing strategy aimed at repositioning the sector in relation to its rivals in Japan and Hong Kong. Amy K. Glasmeier showed that the reason the Swiss watchmakers regained their ability to compete was the re-adaptation of the Swiss production system to global competition during the 1980s – what she calls "reorganization à la Japanese,"[7] or mass production of a limited number of mechanical calibres (movement without regulating parts), which was the strategy adopted by Hattori & Co. (Seiko) in the 1950s. The work done by Olivier Crevoisier and his team not only highlights the transformation of production systems but also underscores the decisive role played by the new marketing strategies. According to Hugues Jeannerat and Olivier Crevoisier, the success of the Swiss watchmaking industry was even due to precisely "non-technological innovations."[8] Leila Kebir and Olivier Crevoisier clearly showed how the return of mechanical watchmaking in the 1990s was based on the use of such cultural resources as "historical legacy, technique and aesthetics."[9] This process of building a cultural heritage also figures prominently in the work of the ethnologist Hervé Munz.[10] Contrary to the cultural school, it is not the existence as such of traditional know-how inscribed in the territory, but rather the use and maintenance of this image for marketing purposes which matters. However, this regional economic research primarily covers the period after 2000 and fails to position this new strategy within the context of a historical process, which could shed light on its origins.

This book adopts the latter approach, but also attaches real importance to an analysis of how the production system is organized. To grasp how Switzerland has been able to return to dominance of world watch markets, one must distance oneself from the high-flown narratives on the age-old know-how and technical excellence of Swiss watchmakers. The circulation of technologies and know-how worldwide is a general phenomenon which has been accelerating since the Industrial Revolution and affects all sectors of the economy.[11] Nowadays, high-quality mechanical watches can be manufactured anywhere in the world. Seiko has been making since the 1960's extremely accurate high-quality mechanical watches and has the know-how to produce *tourbillon* watches, which are considered the pinnacle of the watchmaking profession (the tourbillion is a small device which eliminates errors of rate in the vertical positions).[12] The fact that it has not done so, despite holding the necessary patents, is due to marketing reasons, not technical ones: on the global luxury goods market, watchmaking is Swiss by its very nature. Since the mid-1990s, the transformation of Swiss watchmaking into a luxury industry and its adaptation to the sweeping changes in this sector appear to be the very basis of its success. This far-reaching change is particularly visible when we look at statistics for watch exports.

General trends in Swiss watch exports

Swiss foreign trade statistics illustrate the overall trend in the Swiss watchmaking industry between 1985 and 2010 (see Table 1.1). There were two distinct phases. First, the years 1985–1995 were the decade when the industry overcame the crisis, as reflected in the overall upturn in exports (in terms of both volume and value) and a sharp rise in the production of quartz watches. The latter accounted on average for 89.2 per cent of the volume of watch exports in the 1990s but barely half of their value. These figures highlight the importance of the restructuring of production systems and efforts to compete with Japanese rivals during this phase.

Second, the years 1995–2010 saw the introduction of a marketing strategy which turned Swiss watches into luxury goods. The volume of exports declined sharply, stabilizing at around 30 million pieces in the second half of the 2000s. On the other hand, the value of exports posted extraordinary growth, reaching CHF15.3 billion in 2010. This boom was based on the strategy of turning mechanical watches into products of prestige, tradition, and luxury: their share of the total export value shot up from 47.3 per cent in 1995 to 71.9 per cent in 2010, whereas their volume showed only a slight rise. In 2010, more than 80 per cent of all Swiss watch exports were quartz watches. Since 1995, however, growth in the watchmaking industry has been driven by the sharp increase in the price of mechanical watches, which are now marketed as luxury goods. Yet, even though marketing strategy appears to have been a decisive factor in this process, production-related aspects must also be taken into consideration.

Table 1.1 Swiss watch and movement exports, 1980–2010

	1980	1985	1990	1995	2000	2005	2010
Volume (millions of pieces)	47.1	37.4	41.6	43.7	35.9	29.9	31.9
Mechanical (%)	79.3	19.3	9.5	13.0	9.7	14.2	19.2
Quartz (%)	20.7	80.7	90.5	87.0	90.3	85.8	80.8
Value (CHF millions)	2,790.1	3,595.8	5,936.2	6,793.5	9,402.5	11,560.7	15,343.0
Mechanical (%)	68.4	43.4	41.5	47.3	47.5	62.2	71.9
Quartz (%)	31.6	56.6	58.5	52.7	52.5	37.8	28.1
Average mechanical value (CHF)	51.0	216.0	623.5	567.4	1,285.9	1,694.6	1,805.5
Average quartz value (CHF)	90.3	67.5	92.2	94.2	152.2	170.6	167.3

Source: Federation of the Swiss Watch Industry.

The globalization of the luxury industry

In order to understand the circumstances in which the Swiss watch industry has staged a triumphant comeback on world markets since the late 1980s, as well as the Swatch Group's strong growth, we must look at the development of the luxury industry as a whole. Watchmaking has been affected by the sweeping changes which have impacted all the industries in this sector since the 1970s, including fashion, wine, and jewellery. The transformation of the luxury industry has been characterized by the rationalization of production, the globalization of brands, and the democratization of use.

Traditionally, luxury products were goods produced in small quantities, often by small family firms, and meant for a small, wealthy elite. However, the creation of multinational companies in the luxury industry, often based on the acquisition and merger of former family firms, has profoundly changed the nature of the products: luxury goods have become objects of mass consumption, bearing hefty price tags but being produced industrially and meant for both the middle and upper classes. Management researchers have developed various concepts to describe the emergence of these new consumer goods. For example, Danielle Allères speaks of "affordable luxury,"[13] while Michael J. Silverstein and Neil Fiske use the notion of "New Luxury."[14] The production and sale of such goods has become an extremely profitable activity as a result of the large profit margins generated by an exclusive image maintained by heavy advertising. This can be contrasted with traditional luxury, which is categorized as "exclusive," involves extremely expensive goods, targets wealthy customers, and usually generates smaller margins.

This transformation has attracted a great deal of attention from management researchers, spinning off a new field of research peculiar to this discipline. Yet these efforts appear insufficient to account for the transformation of Swiss watchmaking in all its complexity. Likewise, modern developments in the European luxury industry remain relatively unexplored by economic historians. The work done in management and the social sciences is generally characterized by a lack of historical perspective and distance in relation to the brands' version of their own history. A great deal of marketing and management research concentrating on the luxury industry is published by researchers who work as consultants for the large firms in the sector, for which they train the new generations of managers in their universities. This type of research, which is widespread, is aimed more at providing an academic justification for the brands' narratives than at furthering understanding of their marketing strategies. To cite just one example, Jonas and Bettina Hoffmann wrapped up a study on the watchmaker Richard Mille by writing that "passion is a distinctive characteristic of luxury innovators in their path; it is the source of insights; and it is passion in the execution and the delivery of a rare experience that will make luxury innovation succeed and sustain brand equity in the long run."[15] Marie-Claude Sicard accurately sums up the dominant trend in this field: "Owing to a surprising consensus, no one – neither the media, nor the analysts, nor the authors writing on a given subject – dares challenge conventional wisdom. Stereotyped language prevails, whose only purpose seems to be mutual reinforcement of each other's advertising supports."[16] What is more, the historians of the luxury industry have tended to focus on the 18th and 19th centuries. In the cases of France and Great Britain, they show how luxury goods, which were originally meant for the aristocracy, flourished after the urban middle classes emerged and production was mechanized. However, their research tends to stop at the 1970s.[17]

Yet the threefold trend towards rationalization, globalization, and democratization which has characterized the luxury industry – and watchmaking in particular – in recent decades appears to be a genuine shift. It is based on a transformation in the nature of products, with a shift from high-quality goods manufactured in a quasi-traditional manner, often by family firms ("exclusive luxury," such as limited-series watches sold for several hundred thousand francs a piece), to objects of mass consumption identified by global brands ("affordable luxury," such as Omega, TAG Heuer, and Rolex). This sea change has caught the attention of several historians, primarily in France and Italy, who have demonstrated that the boom in fashion, jewellery, and foodstuffs has been driven since the 1970s by the adoption of new, globally conceived marketing strategies (retail networks, branding policy, design, etc.) and the grouping together of former family and crafts firms within luxury groups such as LVMH (Moët Hennessy Louis Vuitton), which was founded in 1987 by French businessman Bernard Arnault. However, their work focuses on creation, consumption, and marketing, leaving production issues almost entirely aside. When an author refers to "the industrialization of luxury," it is primarily to highlight, often with regret, the fact that use of luxury products has become democratized and commonplace – not to explain the productive foundations of this industry. Corinne Degoutte is one of the few who mention the challenge posed by the redefinition of "production strategies," but this topic remains secondary in her argument.[18] Andrea Colli and Elisabetta Merlo show that the success of the Italian fashion industry from the 1970s onwards has been based on the partnership between independent designers and small and medium-sized enterprises from the industrial districts of leather goods or textiles, but they fail to look at how this production system has evolved.[19]

However, it appears highly unlikely that the competitiveness of the European luxury industry is entirely based on image, brand, and communication – in short, *software*. Clearly, the democratization and rapid expansion of the "New Luxury" brands has been accompanied by a restructuring of production systems which allows these companies to offer products which not only match consumers' tastes but also compete on cost. Production systems and marketing strategies appear to be the two complementary pillars of the modern luxury industry.

This is particularly the case for the Swiss watch industry and the Swatch Group. It would appear that Switzerland's restored ability to compete with Japan is precisely due to a new marketing strategy which turned watches into "affordable luxury" goods, combined with a thorough restructuring of production systems. And it is precisely this dual strategy that is the subject of this book, which follows a structure that is both chronological and thematic, and consists of ten chapters. Chapter 2 looks back at the watchmaking crisis of 1975–1985 and the reasons the Swiss watch industry was having difficulty competing with its main rival, Japan. The creation of the Swatch Group, which is the subject of Chapter 3, was the main measure introduced in the early 1980s to overcome the crisis. The restructuring of production systems subsequent to the company's founding was accompanied by the launch of an emblematic new product on world markets: the Swatch. This chapter also looks at the impact of this famous plastic watch on corporate management. The next two chapters focus on the two lynchpins of the new strategy adopted by Nicolas G. Hayek as the head of the Swatch Group up to the late 1990s: the rationalization and globalization of production systems (Chapter 4) and the adoption of a new marketing strategy (Chapter 5). As these changes were being implemented, the Swatch Group began in the mid-1990s to transition slowly but surely to luxury, and the broad lines of this transition are sketched out in Chapter 6. Two elements at the heart of this repositioning strategy are analysed in depth in the following two chapters: the rebranding of Omega, which became the Swatch Group's main brand, positioned in the profitable New Luxury segment (Chapter 7); and the Chinese market, which in the 2000s became the Group's number one customer (Chapter 8). Chapter 9 deals with the Swatch Group's main rivals, both in Switzerland and elsewhere, in order to highlight the Group's position as the leading watchmaker on world markets. Chapter 10 concludes.

2
The Watchmaking Crisis of 1975–1985

During the summer of 2010, the Swiss press and international media paid a vibrant tribute to Nicolas G. Hayek, who passed away on 28 June 2010 and was acclaimed as the saviour of the Swiss watch industry. The American business magazine *Forbes* celebrated "a legend... credited with engineering the rebirth of the Swiss watch industry",[1] while *The New York Times* wrote about "a flamboyant figure [who] saved the Swiss watch industry with the introduction of the Swatch."[2] Indeed, when the big Swiss banks hired Hayek as a consultant in the early 1980s, the Swiss watch industry was in the grip of an economic crisis which threatened its survival. After nearly 40 years of virtually steady growth (the drop in exports of watches and movements during World War II had been more than offset by the production of munitions for the belligerents), the Swiss watch industry entered a recessionary phase in 1975. Swiss foreign trade statistics clearly reflect this phenomenon (see Figure 2.1). The volume of exports soared from 24.2 million watches and movements in 1950 to 40.9 million in 1960 and peaked at the historic level of 84.4 million in 1974, before falling to an annual average of 31.3 million pieces in 1982–1984. As for the number of people employed, it nosedived from almost 90,000 in 1970 to fewer than 47,000 in 1980.

Despite the impact and scope of this crisis in the history of the Swiss watch industry, paradoxically enough, it did not generate academic works in business history or in economics that would make it possible to consider the origins, mechanisms, and consequences of the crisis. Yet we must examine the precise causes of the 1978–1985

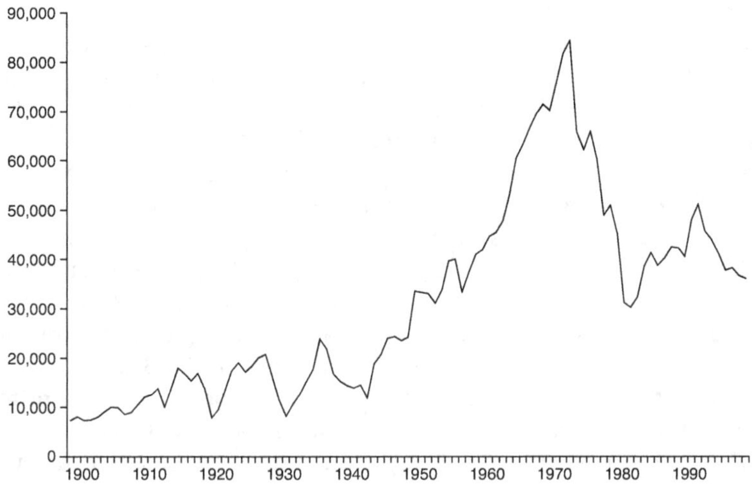

Figure 2.1 Swiss exports of watches and movements, volume as thousands of pieces, 1900–2000

Source: Swiss foreign trade statistics, Bern: Federal Customs Department, 1900–2000.

crisis in the Swiss watch industry to understand the issues at stake behind the two-pronged strategy of rationalizing production systems and rethinking marketing, both of which were adopted following the creation of the Swatch Group.

The 1975–1985 crisis in historiography

Most of the authors who tackled the 1975–1985 crisis in the Swiss watch industry highlighted a group of factors (a lack of industrial concentration, the quartz revolution, the strong Swiss franc, the oil price shock, etc.) as explanatory elements without showing how they hinged upon each other and led to a crisis, as if their simultaneity sufficed to make them explicative factors. Two works published in the 1980s took up the question of the "watch crisis". They largely inspired the authors who published on this subject in the 1990s and the 2000s.

The first is the book by Georges Piotet, the outcome of a PhD thesis in social sciences written at the University of Lausanne, in which he analysed structural changes in the Swiss watch industry from the angle of corporatism.[3] In a first part devoted to the "period of problems" (pp. 37–70), he emphasized the difficulties encountered by the watch industry at the beginning of the 1980s. Five factors were highlighted: the industrial structure (numerous SMEs), the technological revolution (quartz watch), international competition, currency instability, and a changing economic climate. Yet the author did not offer the reader any general, systemic interpretation of the crisis, limiting himself to successive descriptions of the "main problems encountered that capital, the trade unions and the State tried to solve".[4]

The second is the consolidated book on the global history of horology and time measurement, published in 1983 by Harvard University professor David S. Landes. The author devoted an entire chapter to the 1975–1985 crisis in the Swiss watch industry, entitled "The Quartz Revolution".[5] He tackled the question of new competitors who challenged Switzerland's domination of world markets after World War II. He presented the case of the American company Timex Co. and its system for mass producing bottom-end mechanical watches, which in the end did not appear as an existential threat to Swiss watchmakers, for unexplained reasons. In the end, according to this author, it was a product innovation, that is, the quartz watch, which precipitated the crisis. Landes spoke of "a radical transformation of the technology of time measurement [which] resulted in the creation of what still looked like a watch but was in reality a new product."[6] Going on to adopt a classical perspective of the history of techniques, he showed how some prototypes were developed in Switzerland, in the United States, and in Japan in the late 1960s. Yet despite its mastery of this new technology, the Swiss watch industry did not embark upon the production of quartz watches: "the only thing lacking was entrepreneurship: the manufacturers of watches were not interested."[7] According to Landes, the inability of Swiss watchmakers to adopt new technologies led in this way to the 1975–1985 crisis, the closure of companies, and a drop in employment.

The works published in the following decades that dealt with the question of the 1975–1985 crisis were largely inspired by the works of Piotet and Landes.[8] They emphasized the inability of the Swiss watch industry to adopt this major technological innovation and viewed this as the primary cause of the crisis, which was supposedly aggravated by the strength of the Swiss franc, a link which, however, has not been proved. The main problem with these studies, as with the publications of Piotet and Landes, is that they did not construct any model explaining the crisis but instead presented some elements primarily related through their simultaneity. In fact, the academic debate on the 1975–1985 watch crisis very largely reflects contemporary actors' opinion of their own action, thus transmitting an image similar to the one shared in the Swiss collective memory. In this respect, the work of Cécile Aguillaume on the crisis as viewed through the regional and national press is highly representative of a perspective which reproduced the narrative of actors as an interpretative factor rather than reconstructing the mechanism of the crisis.[9] Landes also relied essentially on the press reports and the testimony of entrepreneurs and engineers. This traditional image of a major crisis due to the inability of the Swiss watch industry to cope with the quartz revolution is, moreover, reinforced by the quasi-legend that has arisen with regard to Swatch and Nicolas G. Hayek, taken up by historians.[10] According to this narrative, it was a providential man, coming from the outside, who rebooted the industry with the help of a new product that conservative manufacturers did not want to produce: a quartz watch made out of plastic. Even if there is a grain of truth in this story, it obscures an essential feature of the regained competitiveness of the Swiss watch industry: the deep restructuring of its production system. More than a product innovation, a process innovation was indeed at the heart of the lost competitiveness of the Swiss watch industry in the years of crisis between 1975 and 1985.

A technological challenge: the mass production of high-quality mechanical watches

The crisis of 1975–1985 was indeed due to the structural unsuitability of the Swiss watch industry to acquire new technologies and integrate them into the production system. However, this was not originally a problem with the conception and manufacturing of a new product (quartz watch), but rather with the organization of the mass production system. While the literature largely focused on innovation-related questions, the authors strangely enough failed to distinguish between *product innovation* and *process innovation* – a distinction which goes back to Schumpeter[11] – focusing only on the former.

Thus, the overemphasis on objects led historians to hold the so-called quartz revolution responsible for the 1975–1985 crisis. Yet a comparative analysis of the Japanese competitor reveals a more complex process. On the one hand, the fact that Seiko was the first watch company to launch a quartz watch on the market, at Christmas in 1969, does not reflect a technological advantage over its Swiss rivals (Longines, Omega, Ebauches). Rather, it was the fruit of a marketing strategy aimed precisely at establishing its image of a leader in matters of technology and precision.[12] At the time, precision was indeed the key point of Swiss watchmakers' communication strategy. In any case, the Swiss companies soon followed Seiko's lead, launching their first quartz watches (Girard-Perregaux in 1971, Longines in 1972, Omega in 1973).[13] As a result, the question of primacy and backwardness has no bearing on technological mastery. On the other hand, even if they did promote quartz watches, the Japanese watch companies only gradually shifted to this new product.

Figure 2.2, which represents Japanese national production of both mechanical and quartz watches, brings forth two comments. First, production of quartz watches began to increase in the second half of the 1970s. In 1975, it amounted to only 2.6 million pieces, that is, 8.5 per cent of the total. Moreover, mechanical watches still accounted for over half of overall watch production until 1978. Only in 1979 did quartz watches overtake mechanical watches in terms of national production, with 33.9 million pieces and 57.0 per cent of volume. Afterwards, the pace of growth was very fast, with production of quartz watches amounting to 53.8 million pieces (62.4 per cent) in 1980 and 149.3 million (81.7 per cent) in 1985. Second, the Japanese watch companies did not stop producing mechanical watches even after they had based their growth on quartz watches. Of course, the production of mechanical watches did not continue to grow at such a fast rate as during the 1950s and 1960s, but there was no sharp drop after 1975, as could be seen in Switzerland. Mechanical watches went from 23.8 million pieces in 1970 to 25.6 million in 1980, peaking at 41.8 million in 1986. There was still a market for mechanical watches, and the Japanese were taking it from the Swiss. As a conclusion, it must be admitted that the real shift towards quartz watches occurred in Japan after the crisis had begun in Switzerland. The 1975 drop in Swiss watch exports cannot be directly linked to the quartz

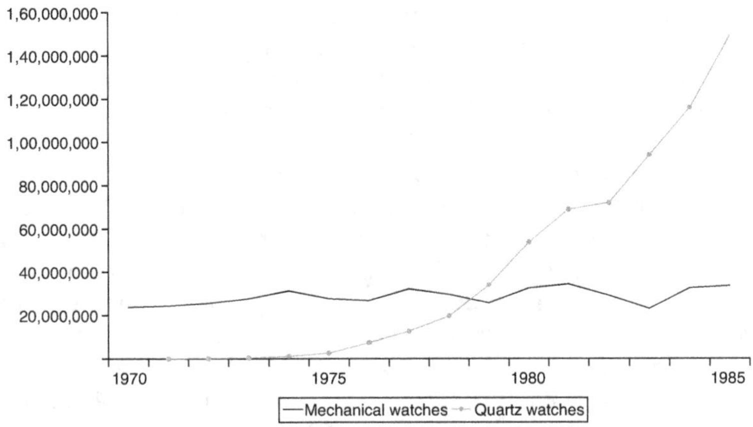

Figure 2.2 National production of watches in Japan, volume, 1970–1985
Note: estimates for quartz watches in 1971–1974 are based on an extrapolation of Seiko data.
Source: *Nihon tokei nenkan*, 1970–1985.

revolution – even if it reinforced Japan's competitive advantage in the early 1980s – and the crisis in Switzerland.

Rather, we must look for the origins of the crisis in Switzerland in the problem of process innovation. Mechanical watch production indeed underwent a far-reaching change in the 1950s and the 1960s with the adoption of the mass production system, which was only partially implemented in Switzerland. Until the mid-1980s, a growing bipolarization could be seen in the Swiss watch industry. On the one hand, quality watches were still manufactured within an unrationalized production system. Strengthened and supported by the legislation on the watch cartel, the so-called *Statut horloger* (1934–1971), the dispersion of the industry between numerous independent companies precluded a rationalization of the production of parts and movements.[14] This was above all the case with the *manufactures*, which were legally obliged to produce their own calibres themselves. In 1936, they numbered 48 and represented the most renowned names in the industry (Longines, Nardin, Omega, Patek Philippe, Rolex, Tissot, Vacheron Constantin, etc.).[15] This lack of rationalization was also due to the existence of many independent firms and the virtual absence of corporate groups, as a result of which each company tried to be active in all the segments of the market and offered a wide range of watch types under its own brand. At the beginning of the 1980s, Omega is said to have had more than 1,600 different models.[16] Even if this extreme case may not be representative of the industry as a whole, it underscores the impact of the absence of marketing rationalization on production.

On the other hand, a mass production system was being introduced for the manufacture of cheap and simple watches (pin-lever watches, also called *roskopf* watches). Made in Switzerland from the end of the 19th century onwards,[17] they recorded high growth worldwide after World War II thanks to the introduction of mass production methods. The saga began in the United States with a former munitions manufacturer, Timex Inc. Founded in 1941 by two Norwegian immigrants, Thomas Olsen and Joakim Lehmkuhl, who bought up the Waterbury Clock Company, Timex produced munitions during the war, then transitioned in 1949 to the mass production of cheap wristwatches.[18] As a result, the production volume of Timex watches grew from 1 million pieces in 1949 to 8 million in 1960 and 22 million in 1969.[19] They occupied the bottom end of the range until the late 1970s, when they were replaced by cheaper quartz watches. Competition from Timex on the world market led some Swiss companies to embark upon the production of pin-lever watches. This was particularly the case with Baumgartner Frères SA (BFG) in Grenchen.[20] BFG became one of the main Swiss watch companies and opened a subsidiary in Hong Kong in 1970 for the production of pin-lever watches.[21] As for the Société suisse pour l'industrie horlogère (SSIH, grouping together notably Omega and Tissot), in 1971 it acquired a majority stake in Economic Swiss Time Holding, a group founded in 1967 to compete with Timex.

Thus, although Swiss watch companies had mastered the technologies for the mass production of watches in the 1960s, with the exception of the Société des Montres Rolex SA, they did not adapt them to the manufacture of high-quality goods. There was a real gap between quality watches, where production was not streamlined, and cheap, mass-produced watches, with both production systems coexisting sometimes within a single company. SSIH is a good example of this duality: in 1973, pin-lever watches represented 69.6 per cent of its sales volume but only 18.9 per cent in terms of turnover.[22] Yet it was precisely by merging these two models, that is, by mass producing quality watches, that Japanese watchmakers were able to establish themselves on world markets as the challengers to Swiss watchmakers.

Competition with Japan

A quick comparison between Omega and Seiko, the biggest watch companies in Switzerland and Japan, respectively, highlights the existence of fundamentally different production systems (Figure 2.3). In terms of production volume, Seiko overtook its Swiss rival as early as 1949, and its production was more than double Omega's from 1953. However, it was especially in the second half of the 1950s that the Japanese firm experienced dramatic growth and reached a production volume of 3.7 million pieces in 1960 and 14.0 million in 1970. Unlike Omega's, this rise was based not on an increase in the number of models but rather on the launch of a very limited number of models that were selected for mass production. Thus, between 1945 and 1967, Seiko marketed only 123 different models – mostly

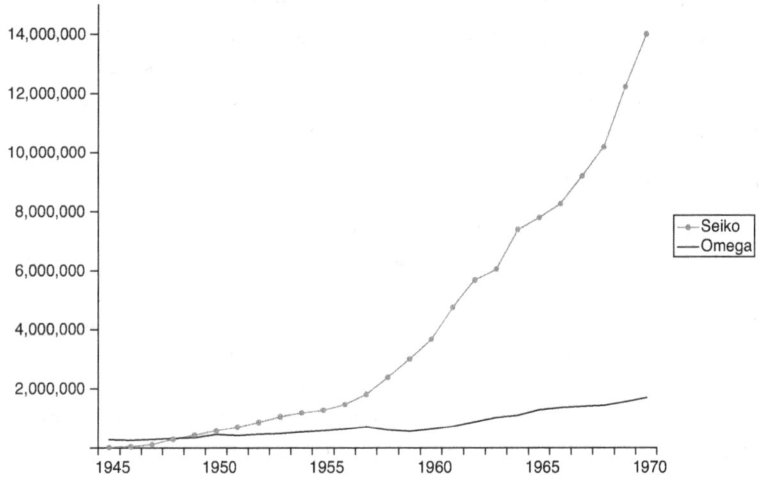

Figure 2.3 Production of watches by Seiko and Omega, volume, 1945–1970

Source: For Omega, Pasquier, *La "recherche et développement" en horlogerie*: 440–441; for Seiko, statistics communicated by the Seiko Institute of Horology, Tokyo.

improvements of previous models – as compared with Omega's several hundred.[23] Moreover, these were not low-end watches. Only in 1971 did Seiko open a factory for the production of pin-lever watches with the aim of checking the inflow of cheap Swiss and American watches to the Japanese domestic market.[24] Seiko's engineers rather developed high-quality goods. In 1959, they launched the first self-winding watch, which became the top-end watch after the war,[25] and went on to mass produce it: the production volume of self-winding watches manufactured by the company Suwa Seiko – one of the two watch companies of the Seiko group – soared from 430,000 pieces in 1961 to nearly 4.3 million in 1970.[26] Moreover, the Japanese firm introduced a mass production system at all its plants, symbolized by the incorporation of conveyor belts into assembly lines in 1958.[27] The triumph of Seiko's mechanical watches at the competitions of the observatories of Neuchâtel (1967) and Geneva (1968) illustrated the success of the production system adopted by Japanese watchmakers.[28] Finally, this streamlining of production also made it possible to follow a highly rational marketing strategy organized on a global scale (few models, high-density sales network, and mass advertising). In this way, high-quality mass-produced watches enabled Japan to challenge Switzerland's domination of the world market in the 1960s and the 1970s. Afterwards, from the end of the 1970s onwards, product innovation, in the form of quartz watches, reinforced this competitive advantage. The examples of Hong Kong and the United States clearly illustrate this phenomenon.

The first clash: Hong Kong

As a commercial hub in Asia, Hong Kong was one of the first places where Japanese watchmakers undertook their worldwide expansion in the early 1960s (see figure 2.4). At the time, the watch market was controlled by the Swiss, who had not yet encountered Japanese competitors. Indeed, the situation of the Swiss watch industry in Hong Kong was a virtual monopoly: imports of watches and clocks from Switzerland amounted to 89.1 per cent of total value in 1960 and to 75.8 per cent in 1965.[29] Yet Swiss watchmakers gradually lost market share as the Japanese penetrated this market. Even though imports from Switzerland were still 11.8 times greater than imports from Japan in 1964, this difference fell to 5.7 times in 1965 and to less than 2 in 1968. Finally, in 1975, the decline in imports from Switzerland made

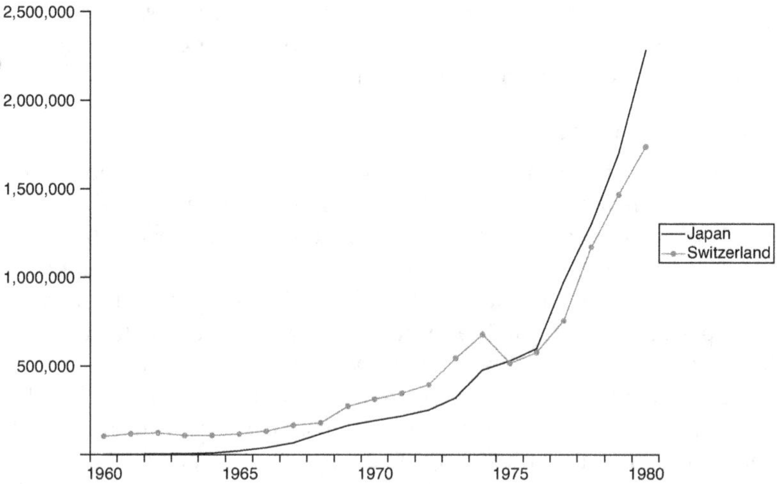

Figure 2.4 Imports of watches and clocks into Hong Kong, in HKD thousands, 1960–1980

Source: Hong Kong Census and Statistics Department, *Hong Kong Trade Statistics. Imports*, 1960–1980.

it possible for Japan to overtake its rival. Afterwards, both countries experienced growth but Japan had outstripped Switzerland.

Initially, Japan's fast growth was not due to quartz watches. Hong Kong's foreign trade statistics for 1975 do not distinguish between watch types, but it seems evident anyway that electronic watches accounted for only a small part of imports from Japan – they amounted to a meagre 8.5 per cent of Japanese domestic watch production and were quite expensive products.[30] If one considers only finished watches, in 1975 Japan's share amounted to HKD278.4 million as compared with 259.4 for Switzerland.[31] In the second half of the 1970s, Japan beefed up its position with the help of electronic watches. In 1980, Hong Kong imported finished watches from Japan to the tune of HKD785.5 million, of which 64.7 per cent were electronic watches, while Switzerland's share was HKD695.6 million, of which only 3.8 per cent were electronic watches.[32] Thus, even on a market such as Hong Kong where electronic watches clocked up steady growth, Japan's success could be traced to mechanical watches.

They made possible the commercial expansion, which was then reinforced by quartz watches. The situation was similar elsewhere in the world, in particular in the United States.

The American market

The American watch market perfectly highlights the two-stage expansion of the Japanese watch industry. The US was the world's largest market in the second half of the 20th century. Moreover, the Swiss watch industry relied heavily on it, despite the rocky commercial relations between the two countries:[33] the American outlet absorbed 35.5 per cent of all watch exports from Switzerland in the 1950s.[34] Thus, the arrival of Japanese competitors posed a serious threat, which in fact was not limited to this market alone but applied to the industry as a whole. Until the mid-1960s, Swiss watchmakers had a monopoly on this market, with 92.2 per cent of imports of complete watches to the United States in 1965. Japan was nearly absent: that same year, American imports of Japanese watches came to a paltry USD74,000, that is, 0.2 per cent of the total.[35] However, Japanese watchmakers were present on the market through assembly factories set up in the Virgin Islands that belonged to American watchmakers and distributors, to whom they supplied watch movements sourced from Japan.[36]

The Japanese watch companies organized their arrival on the American market in the second half of the 1960s.[37] Watch imports from Japan increased sharply in comparison with Swiss watches, where growth was much slower (see Figure 2.5). In value terms, Japanese watches amounted to USD2.8 million in 1970, that is, 4.0 per cent of the total, as against less than 1 per cent up to 1968. This fast growth continued in the years 1970–1975, when the value of watch imports from Japan increased nearly tenfold, reaching USD26.5 million and representing 17.6 per cent of total imports in 1975. Yet even if American statistics do not allow us to measure this precisely, mechanical watches still accounted for the bulk of imports, as in Hong Kong. This growth was driven by the strong competitiveness of Japanese watches in terms of price and quality, a perfectly planned marketing strategy (advertisement, distribution network, after-sales service) and a favourable exchange rate (see below). Swiss watchmakers soon lost market share to these competitors. Even though imports of Swiss watches continued to increase in absolute terms despite the economic downturn of

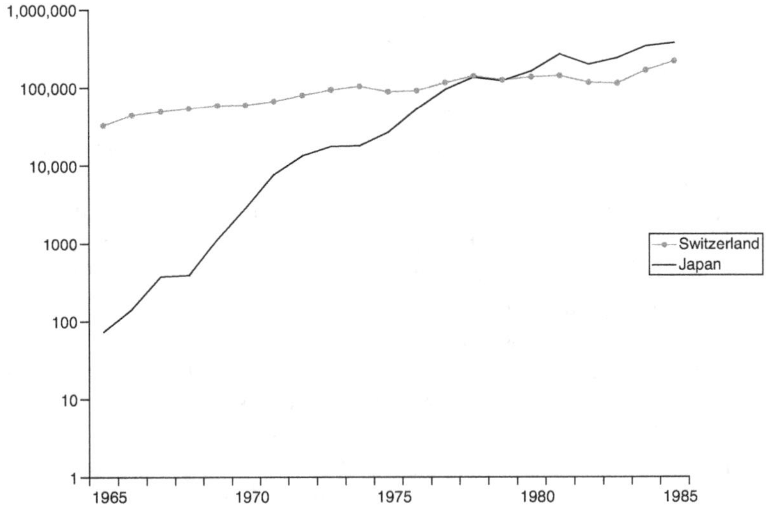

Figure 2.5 Imports of complete watches into the United States, in USD thousands, 1965–1985 (logarithmic scale)

Source: U.S. Department of Commerce, Bureau of the Census, *US import commodities by country*, 1965–1985.

1975–1976, the flow was far lighter than it was for Japan. Between 1970 and 1975, the value of imports of Swiss watches did not even increase twofold, rising only from USD59.6 million to 88.4 million. Moreover, market share plummeted, declining from 83.1 per cent of total imports in 1970 to 58.8 per cent in 1975.

As a result of this trend, which was favourable to Japan, the country experienced strong growth during the following decade. Reliance on marketing facilities set up in 1965–1975 led to massive sales of quartz watches. Imports of Japanese watches came to USD168.2 million in 1980 and 373.4 million in 1985. Japan overtook Switzerland in 1980, when its market share hit 26.5 per cent, then forged ahead with 36.8 per cent in 1985. During this same period, Switzerland entered a phase of stagnation, with an average US import value of USD127 million for the years 1977–1983 and a market share which dropped to 22.2 per cent in 1980, then sank further to a low of 15.3 per cent in 1983. The launch of the Swatch, which was a huge hit in the United States, clearly explains the comeback of 1984–1985.

The impact of monetary liberalization

There was a final element which historians of the watch industry have hardly tackled in their explanation of the watch crisis but which was nonetheless a key element: the monetary factor. Analyses usually emphasized the problem – incidentally omnipresent in archives – but failed to integrate it into a general explanation of the crisis. A comparative analysis of exchange rate trends for the Swiss franc and the Japanese yen against the US dollar in the 1970s and 1980s helps us understand better the unfavourable impact of the monetary factor for Swiss watchmakers (Figure 2.6).

The international monetary system experienced a deep shift in 1973 with the end of the fixed currency system and the introduction of a floating exchange rate system. As far as competitiveness was concerned, this change had important consequences for the watch business because it caused the Swiss franc to appreciate sharply, raising the cost of Swiss watches on the world market, particularly in the US, precisely at a time when Japanese competitors were growing

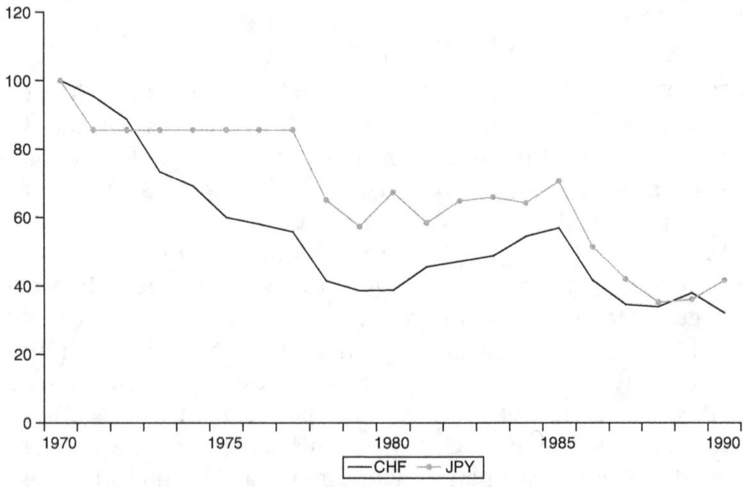

Figure 2.6 Exchange rate trends for the Swiss franc (CHF) and the Japanese yen (JPY) against the US dollar (1970 = 100), 1970–1990

Source: For CHF, statistics of the Swiss National Bank (www.snb.ch, accessed 7 July 2010); for JPY, *Nihon chouki toukei souran*, Tokyo: Nihon tokei kyokai, 1988, 18–8.

stronger. Between 1970 and 1979, the dollar lost more than 60 per cent of its value against the Swiss franc. At the same time, however, Japan continued to benefit from a fixed exchange rate with the United States until 1977, although the yen was revalued from 360 (its fixed parity since 1949) to 308 yen for 1 dollar in 1971. Obviously, this stability is the main reason why Japanese watchmakers did not suffer unduly as a result of the oil price shock: indeed, American imports of Japanese watches grew from USD17.9 million in 1974 to 26.5 million in 1975 and 52.9 million in 1976.[38] Afterwards, even though the dollar lost ground against the yen from 1977, the drop was less brutal than for the Swiss franc. Until 1985, the difference between the currencies against the US dollar was extremely favourable for Japan, thereby sustaining that country's high growth on the American market. Finally, following the Plaza Agreements (1985), the yen firmed up against the dollar and Japan lost a major competitive advantage against Switzerland, especially from 1987 onwards, precisely when the restructured Swiss watch industry undertook its reconquest of world markets, with the Swatch benefiting from an exchange rate that had become favourable in relation to the yen.[39] If one considers the relative size of the American market for the Swiss watch industry, it must be admitted that exchange rate trends are as important a factor as product innovation.

Conclusion

A comparative analysis of the Swiss watch industry's main rival, Japan, brings out the complex linkage between the crisis, production systems, innovation, and exchange rates. Finally, it is possible to assert that the quartz revolution was not at the origin of the crisis but rather heightened its impact. Initially, the adoption of the mass production system for mechanical watches and of a rationalized marketing strategy enabled Japanese companies, especially Seiko, to post strong growth between 1960 and the mid-1970s. The examples of Hong Kong and the US showed that Swiss watchmakers were losing market share to this new rival at the time. Unrationalized production systems and an unfavourable dollar–yen exchange rate were their main weaknesses. In the following decade, the Japanese watch industry grew stronger, due to the quartz revolution and a propitious monetary environment. Quartz watches developed along with very effective organizational capabilities, allowing the Japanese watch industry to keep growing on world markets. Despite the quartz watch boom, however, Japan did not move out of mechanical watches until the mid-1980s. Their production peaked at 41.8 million pieces in 1986 before nosediving (23.5 million in 1990 and 19.0 million in 1995).

3
The Creation of the Swatch Group and the "Swatch Legend"

At the beginning of the 1980s, the main measure adopted by Swiss watchmakers to pull themselves out of recession was the merger of the two biggest Swiss watch groups, namely ASUAG and SSIH (1983). With gross sales amounting respectively to CHF1.3 billion and CHF815 million in 1979, they outstripped the other watchmakers, third place being jointly occupied by the Société des Garde-Temps SA (SGT, brands Avia, Elgin, Fleurier, Invicta, Sandoz, Silvana, Titus, Waltham) and Rolex, whose gross sales at the time were valued at CHF190 million.[1] What is more, in 1979 ASUAG and SSIH employed about half of the workforce in the Swiss watch industry.[2] However, their combined clout must be seen in perspective, as it was primarily due to the crisis which had impacted other companies. In reality, ASUAG and SSIH faced huge industrial and financial difficulties, and their survival owed a great deal to the support of the major banks.

ASUAG, or the need to restructure the production system

ASUAG (Allgemeine Schweizerische Uhrenindustrie AG) was having difficulty managing its growth and diversifying its activities. The company was founded in 1931 to control the production of movement blanks and watch movement parts, in order to limit their export and assembly as finished watches in other countries. This process, known as *chablonnage*, fuelled the expansion of such rivals as Bulova in the United States and Citizen Watch in Japan.[3] ASUAG operated as a holding company which controlled scores of firms and had a virtual monopoly over the production of watch movement parts like movement blanks, springs, and assortments.[4] But its main characteristic was that it reflected the Swiss consensus. The shareholders of ASUAG, whose share capital amounted to CHF10,006,000, consisted of watchmaking manufacturers (CHF5 million), banks (CHF5 million) and the Swiss federal government (CHF6,000 in shares in exchange for the provision of CHF6 million in financial assistance).[5] In addition, the State granted an interest-free loan of CHF7.5 million and banks extended credit lines worth CHF15.5 million, for a total of CHF23 million, which was entirely repaid by the beginning of the 1940s. The federal government justified its financial commitment to a private-sector firm by the fact that ASUAG "could not of course be an ordinary, solely profit-oriented joint stock company. Its task is to safeguard the interests of the watch industry as a whole".[6] The Board of Directors had 30 members, 5 of whom were appointed by the Federal Council. Finally, the Federal Councillor in charge of the Department of Economic Affairs was automatically a member of ASUAG's Executive Management Board, which clearly reflects the degree of importance that the State attached to the company in relation to the watch industry. Thus, from the early 1930s onwards, ASUAG had a monopoly, recognized by the Swiss federal government, on the production of the main spare parts and movement blanks, as watch manufacturers were obliged to buy from ASUAG, except for few companies which produced such parts internally. However, the liberalization of the Swiss watch industry, which was introduced by the Swiss federal authorities during the 1960s in the face of increasing international competition, fundamentally challenged this traditional role of ASUAG, because in theory the production of watch parts became possible for all companies. As a result, the holding company sought to expand its activities to foreign markets. With the ambition of becoming a European giant in the watch industry, it bought up a German movement blank factory, Durowe, in 1964, then acquired a similar plant, SEFEA, in France in 1967.[7] A few years later, in 1971, ASUAG diversified into the production and marketing of finished watches with the establishment of the General Watch Company (GWC), a group which initially consisted of seven watch manufacturers (Certina, Edox, Eterna, Mido, Oris, Rado, and Technos), who were soon joined by Longines and Rotary.

By acquiring factories for producing finished watches, AUSAG hoped not only to secure its sales of movements and watch parts but also to diversify its revenue streams. At the same time, however, there was a trend towards the creation of groups around *manufactures*, which produced their own movement blanks and were thus largely independent of ASUAG. This meant market share losses for the latter, leading it to establish its own watchmaking group.

Initially, this strategy was quite successful. Between 1970 and 1974, ASUAG's turnover posted dramatic growth, reaching CHF1.4 billion in 1974, while it amounted only at CHF 760 in 1970. This expansion was largely due to GWC: watches and finished movements accounted for 43.9 per cent of turnover in 1974 as against a paltry 5 per cent in 1970. What came next was not really a full-blown crisis as far as ASUAG was concerned. Between 1975 and 1982, the holding company went through a phase of stagnation, with turnover averaging CHF1.2 billion, 45.5 per cent of which was generated by the sale of finished watches bearing GWC brands.

Nevertheless, starting in 1975, ASUAG embarked upon a policy of vigorous restructuring with the twofold aim of cutting its production costs and strengthening its ability to compete on world markets. Its movement blanks and watch parts division was not competitive with Asian producers, who were selling very low-cost movements on world markets. What is more, rival groups in Switzerland, such as SSIH, developed their own manufacturing capacity and ordered fewer and fewer movement blanks from ASUAG. Accordingly, there was a need to restructure the production of movement blanks and watch parts within ASUAG, a sweeping project for which the Board of Directors took on Ernst Thomke in 1978. Thomke, a chemical engineer and doctor with knowledge of both mechanics and management, proceeded to implement a policy of merging and streamlining the various subsidiaries of the holding company Ebauches SA (subsequently renamed ETA SA). The total number of ASUAG employees, which topped the 20,000 mark in 1974, declined to fewer than 16,000 in 1980, then fell precipitously to under 8,000 in 1982. In this way, the country's leading watchmaking group demonstrated a real determination to restructure its production system and improve its profitability. Yet these measures alone were not enough and ASUAG had a disastrous year in 1982, when turnover plummeted by nearly 20 per cent.

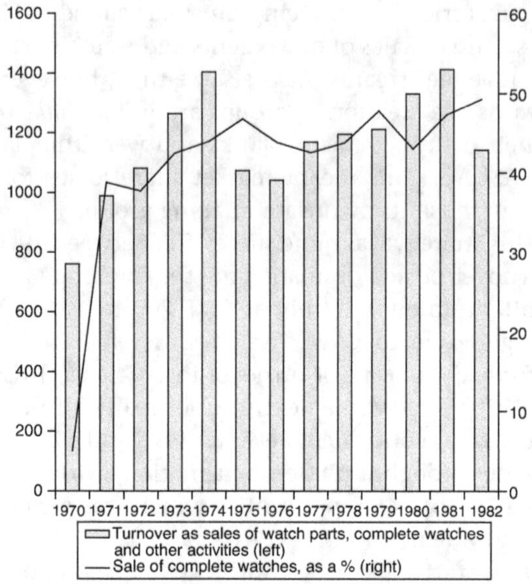

Figure 3.1 Consolidated turnover of ASUAG, in CHF millions, 1970–1982
Source: ASUAG, *Annual Report*, 1970–1982.

SSIH, or the failure of a marketing strategy geared to sales volume

The other major player, SSIH (Société suisse pour l'industrie horlogère SA), Switzerland's second largest watchmaking group, was experiencing severe financial and management difficulties owing to the implementation of a marketing strategy in the late 1960s that was unsuited to developments on world markets. This holding company, which came into being in 1930 following a merger between the firms SA Louis Brandt & Frère – Omega Watch Co, in Biel, and Ch. Tissot & Fils SA, in Le Locle, became the largest watch manufacturing group in Switzerland when it was founded.[8] From the very outset, SSIH based its growth on a policy of buying up other firms, primarily companies specialized in the production of particular types of watch which it did not produce itself. This was the case in particular for the chronograph manufacturer Lémania Watch Co. (A. Lugrin & Co.) in the Vallée de Joux (1932), the ladies' watchmaker Marc Favre &

Co. SA in Biel (1955) and Rayville SA – Montres Blancpain in Villeret (1963), as well as for many small watchmakers scattered throughout the Jura Arc.

As SSIH grew apace, in the mid-1960s it entered the field of so-called "economy" watches, also called *roskopf* or "pin-lever" watches, which were low-quality, mass-produced mechanical watches. Before the advent of quartz watches, this type of inexpensive watch, primarily produced by the American firm United States Time Co. Ltd. (Timex brand) and several Swiss and Hong Kong companies, became very popular around the world. In 1970, they accounted for 43.9 per cent of the total volume of Swiss watch exports.[9] In view of the strong demand for this type of product, SSIH's top management invested massively in this field, backed by such leading Swiss financial establishments as UBS (Union Bank of Switzerland), whose Chairman of the Board, Philippe de Weck, was appointed to the SSIH Board of Directors in 1973. Continuing its policy of acquisition, SSIH bought up successively Langendorf Watch Co. SA (Lanco brand) in 1965, then Aetos Watch SA (1969), followed by the Economic Swiss Time Holding group (1971). These acquisitions significantly boosted the sales volume of the SSIH group, which soared from 3.4 million pieces in 1965 to 4.5 million in 1970, peaking at 13.6 million pieces in 1973, of which 9.5 million, or 69.9 per cent of the total, were pin-lever watches. Turnover also rose steadily, hitting CHF733 million in 1974.

However, this policy of expansion based solely on sales volume rather than efforts to enhance brand value began to sour in the mid-1970s. The arrival on world markets of inexpensive Japanese watches, both high-quality mechanical watches and electronic watches, combined with the sharp appreciation of the Swiss franc, severely constrained the ability of companies like SSIH to compete. As a result, the group entered a time of turmoil, marked by a sharp drop in the volume of sales, which came to a scant 1.9 million pieces in 1980, of which 470,000 were pin-lever watches (which meant that they still accounted for nearly a quarter of total sales). Meanwhile, turnover fell below the CHF600 million mark in 1981. As sales collapsed, SSIH was forced to restructure its companies and make mass redundancies. As a result, staffing declined sharply, falling from a total of 7,300 employees in 1974 to 5,100 in 1980 and fewer than 3,900 in 1982. Finally, the group became increasingly

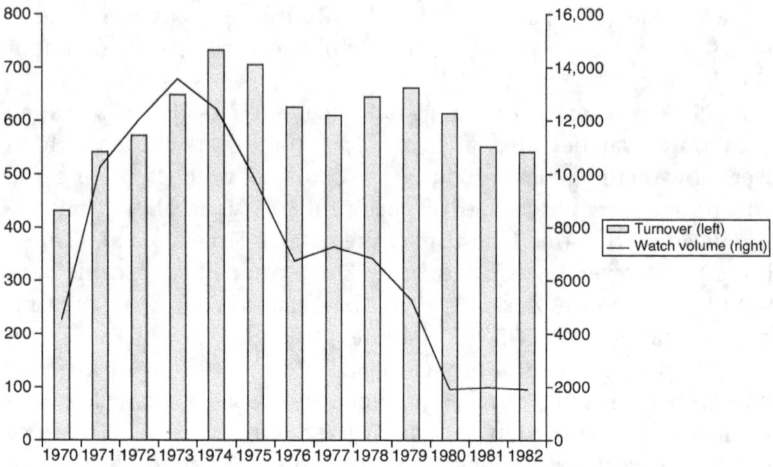

Figure 3.2 Consolidated turnover for SSIH, in CHF millions, and sales volume, in thousands of watches, 1970–1982
Source: SSIH, *Annual Report*, 1970–1982.

dependent on bank capital, as SSIH's share of equity capital plunged from 74.1 per cent in 1970 to a mere 20.7 per cent by 1979. The banking partners, who were keeping the group on life support by granting it medium-term loans, were lobbying hard in favour of its restructuring. A restructuring committee, headed by the Union Bank of Switzerland (UBS), the Swiss Bank Corporation (SBC), and Crédit Suisse, was set up in 1980. The following year, a former UBS director, Peter Gross, took over as Chairman of the Board at SSIH, which was teetering on the brink of bankruptcy.

The founding and beginnings of the Swatch Group (1983–1986)

The dead-end in which ASUAG and SSIH found themselves at the beginning of the 1980s led the Swiss banking partners to turn to consultant Nicolas G. Hayek. Born in Beirut in 1928 and a graduate of the University of Lyon, Hayek founded a consulting firm in Zurich in 1963, Hayek Engineering, which advised industrial companies.[10] In his reports to the banks in 1982, he primarily proposed merging ASUAG and SSIH into a single entity, the Société suisse de microélectronique et horlogerie, or SMH (Swiss Corporation for Microelectronics and Watchmaking Industries Ltd), with the aim of restructuring production systems and adopting a new marketing strategy. This was done in 1983 and gave birth to the world's largest watchmaking group, which went on to take the name "Swatch Group" in 1998.

The years 1983–1986 were a transition period, during which the merger was implemented. The whole structure of the Swatch Group was indeed very unwieldy, as ASUAG and SSIH were both organized as holding companies possessing several groups of enterprises – sometimes with their own subsidiaries. In 1983, the various companies were grouped into three sub-holdings, depending on their type of activity (complete watches; movements and parts; other), the various companies were initially characterized by rationalization. This policy was directed by a four-member Executive Management Board.[11] Chaired by Pierre Arnold, CEO of the Migros chain store and a member of several boards of directors (CFF, Swissair), it included three division managers from the two merged companies (Ernest Thomke for watch production, Andor Helti for other technologies, and Carl M. Meyer for finance). This body worked under the supervision of Hayek, who served until 1986 as a special adviser to the Board of Directors before going on to become the seat of power within the Swatch Group. Hayek became its president, that is, the Swatch Group's CEO, in 1986 and strengthened it in 1988 by appointing eight new members, all active at holding company level (central administration, management, law, finance, real estate, human resources). The Executive Management Board now had 11 members, as against only 4 in 1985. It then took charge of rationalizing the Swatch Group.

As for the Board of Directors, it was still dominated by bankers, who wanted to monitor the success of the transition (see Table 3.1). In 1985, there were four bank directors among the eight members and the position of Chairman was filled by François Milliet, an administrator of several companies including UBS. Arnold, the Swatch Group's CEO, was also a member of the Board, along with two watchmakers, Norbert Schenkel, president of the independent family firm Roventa-Henex SA and a former shareholder of ASUAG, and Paul Lüthi, Chairman of the Board at Rado. This Board did not get involved in the operational management of the company.

Table 3.1 Board of Directors, 1985

Name	Position	Career
François Milliet	Chairman	Executive Chairman of Galenica SA, member of the Board of Directors of the Union Bank of Switzerland (UBS)
Walter G. Frehner	Member	Director General of the Swiss Bank Corporation (SBC), Basel
Pierre Arnold	Member	CEO, Migros
Rolf Beeler	Member	Director General of the Swiss Volksbank, Bern
Peter Gross	Member	Director General of the Union Bank of Switzerland (UBS), Zurich
Paul Lüthi	Member	Chairman of the Board of Rado Watch
Paul Risch	Member	Director General of the Cantonal Bank of Bern, Bern
Norbert Schenkel	Member	Chairman of the Board of Roventa-Henex SA, former Vice-President of ASUAG

Source: Swatch Group, *Annual Report*, 1985.

In 1985, working together with a group of Swiss German investors (Hayek Pool), Hayek acquired a majority stake in SMH. He was then appointed to the Board of Directors, of which he became Chairman in 1986, to replace François Milliet. From that time onwards, he served in a dual position as both Chairman of the Board and CEO until his death in 2010. Shortly after Hayek tightened his grip in this way, two former heads of watchmaking firms, Norbert Schenkel and Paul Lüthi, left the Board of Directors, in 1986 and 1989, respectively. The bankers kept their seats on the Board but had to make room for a new series of investors from Hayel Pool who supported the implementation of Hayek's new strategy (Esther Grether, Peter E. Baumberger, Stephan Schmidheiny, and Franz Wassmer). This transition was essential, because it marked Hayek's arrival at the head of the group and a commitment to his plans to streamline production and adopt a new marketing strategy.

In the short term, the main result of rationalization was that it enabled the company to become profitable once again. After a loss in 1983, the Swatch Group was back in the black the following year. Of course, the profit margin, which amounted to less than 5 per cent of gross sales until 1987, was weak and gross sales grew slowly in the

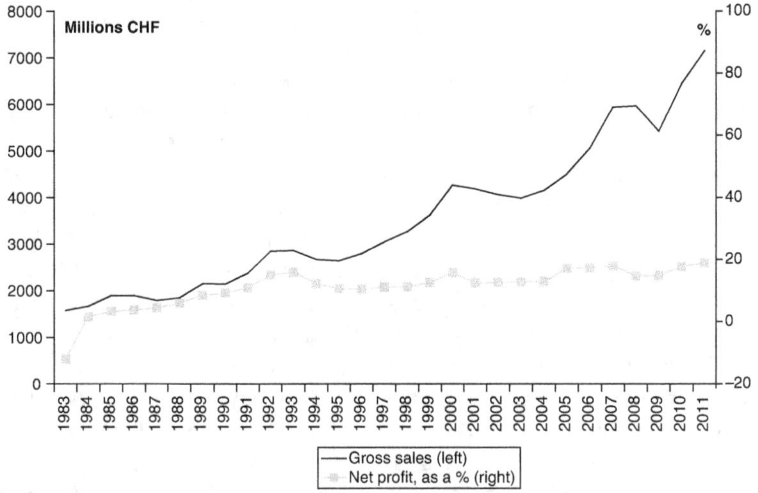

Figure 3.3 Gross sales for the Swatch Group, in CHF millions, and net profit, as a percentage, 1983–2011

Note: Net profit is profit after depreciations, reserves and taxes.

Source: Swatch Group, *Annual Report*, 1983–2011.

1980s. (Incidentally, it was only in 1989 that the Swatch Group hit the CHF2 billion mark in gross sales, that is, the peak reached by ASUAG-SSIH in 1974.[12]) The second half of the 1980s thus appears as the turning point, when the Swatch Group recovered its competitiveness on world markets. Usually, key importance in this process is attributed to the Swatch, but this commonly held view must be examined more closely.

The relative importance of the Swatch

The Swiss collective memory, the story told by the Swatch Group and academic studies usually point to the crucial role of a particular product innovation in explaining the success of the Swatch Group and the comeback of the Swiss watch industry on the world market: the Swatch.[13] Developed by a team of ETA engineers under Thomke's supervision to beat the Japanese competition, the Swatch was a fashion product and a plastic quartz watch manufactured in Switzerland.[14] It was launched in 1983 and experienced growing popularity worldwide from the late 1980s onwards. Its success supposedly enabled the Swatch Group to invest in the take-over and restructuring of other brands, thereby relaunching the entire industry. According to the "Swatch legend", Hayek's launching of a cheap quartz watch, a product which traditional watchmakers did not trust because of their conservative mindsets, made it possible to rescue the Swiss watch industry and enable it to compete with Japan once again.

Due to the lack of data about the various Swatch Group brands and products, it is not possible to assess precisely the Swatch's real impact on the Group's fortunes or determine which products were responsible for its regained competitiveness. However, Swiss foreign trade statistics and the annual reports of the Federation of the Swiss Watch Industry give some valuable clues.[15] They indeed mention the value and volume of exports of "non-metal watches". Even if these watches are not only Swatches, this brand obviously occupies a very important place within this category, with the value of such exports posting high growth after the launch of the Swatch (1983), rising from CHF13.4 million in 1980 to CHF225.9 million in 1985 with a high of CHF798.7 million in 1993 before falling sharply until 2000, then stagnating at an average of CHF297.6 million in 2000–2009, that is, the same level as the second half of the 1980s (see Figure 3.4).

However, these figures do not accurately reflect turnover for the Swatch Group – far from it. First of all, they include products from other brands, even if their share is probably very small owing to the overall tendency to produce metal watches. Second, not all Swatches are included: both watches sold on the Swiss market and metal Swatches are left out. Nevertheless, in the absence of any further data, it is useful to note the value of exports of non-metal watches as a share of Swatch Group turnover.

This comparison highlights an internal trend for the Swatch Group which is similar to the overall tendency. Despite stagnating and then declining slightly in the second half of the 1980s, exports of non-metal watches accounted for a growing share of the Group's turnover, without, however, constituting a majority. For the years 1985–1990, they amounted to only 16 per cent of turnover for the Swatch Group, subsequently rising to an average of 22.8 per cent in the period 1990–1995, with a peak of 28.8 per cent in 1993.

Yet the decline which seemed to occur in the second half of the 1990s was a decline in name only. It affected only non-metal

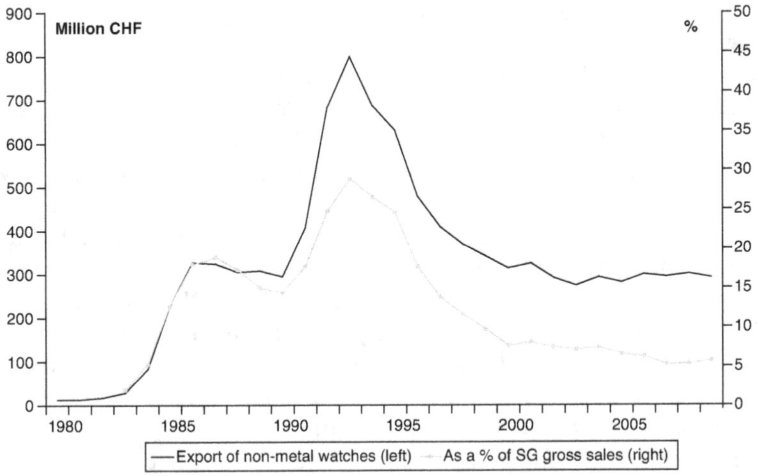

Figure 3.4 Exports of non-metal watches, in CHF millions, 1983–2009
Source: Swiss annual foreign trade statistics, Bern: Federal Customs Department, various years; Swatch Group, *Annual Report*, 1983–2009.

watches, and, coincidentally, Swatch had launched a new collection, Irony, in 1994, composed of metal watches and selling at a slightly higher average price than Swatch plastic watches. According to the "Swatch Finder" from the Internet site for Swatch watches, the share of metal watch models posted strong growth from the late 1990s before levelling off at around one-third in the middle of the 2000s: such watches accounted for 10.5 per cent of new models in 1995; 25.1 per cent in 2000, 38.2 per cent in 2005 and 32.2 per cent in 2010 (www.swatch.com, consulted in March 2012). Thus, the launch of metal Swatches offset to a certain extent the sharp drop in sales of plastic watches. However, overall sales of Swatch watches did not continue on the same level as during the first ten years of the brand's existence. According to the evaluations performed by certain financial establishments for their investor base, total Swatch sales (both plastic and metal watches) came to CHF640 million in 2006 (Helvea SA) and CHF710 million in 2010 (Vontobel). So it is clear that ever since the mid-1990s, the Swatch Group's growth has been powered by other brands in its portfolio.

Consequently, even though it generated large profit margins which gave the company part of the capital it needed for restructuring, the Swatch is more important as a marketing object than in terms of its financial impact. For the first time, the Swatch Group's managers had demonstrated what could be done with a watch: it had become a subject for *storytelling* and major world *events*. Swatch quickly moved into sponsoring extreme sports events for the young, like the Freestyle Ski World Cup in Colorado (1984) or the Impact Tour (BMX and skateboard) in California (1988). This commitment to sponsoring sports events culminated at the 1996 Olympic Games in Atlanta, when Hayek carried the Olympic flame during the opening ceremony. Swatch also commissioned artists to create special Swatch watches, with for example models designed by Kiki Picasso (1985), Keith Haring (1986), and Mimmo Paladino (1989).

But the Swatch's most important innovation was that it was sold as a global brand, that is, a product that is not adapted to local markets but is marketed throughout the world in exactly the same way. Until the 1980s, Swiss watch brands followed a great many different approaches worldwide. Many watch manufacturers exported only naked movements (44 per cent of the total volume of watch exports in 1980),[16] which were encased in Hong Kong or the main markets on which they were sold. This strategy provided a means of cutting production costs, which were high in Switzerland, and offering consumers products that matched local tastes. The design, functions, price, communication, and image attached to brands often varied considerably from one country to another, to such an extent that one cannot speak of a "global brand" during this period. Consequently, the Swatch marked a real break in this field.

This novel marketing strategy was implemented by a new breed of managers, some of whom came from other industrial sectors, especially multinationals dealing in consumer goods, a trend which facilitated the internalization and dissemination of these new practices. Among them, special mention should be made of Max Imgrüth, a graduate of New York's Fashion Institute of Technology, who began his career in the textile industry before joining Omega and taking over product development in the Marketing Department at the end of the 1970s. One of those responsible for launching the Swatch on the US market, he supervised the first test markets in Dallas, Salt Lake City, and San Diego in 1982. He then went on to become CEO of SMH for the United States, one of Swatch's main markets during its phase of expansion. Another name which comes to mind is Jacques Irniger, a former marketing executive for several multinational producers of consumer goods, such as Colgate and Nestlé, who was hired in 1983 to help launch the Swatch.[17] Finally, three years later, the Swatch Group poached Franco Bosisio, a graduate of Bocconi University in Milan, who had worked as a brand manager at Procter & Gamble before becoming marketing director for Cartier Italy. After he was put in charge of the Italian market for the entire Group and appointed marketing director for the Swatch brand in Italy, he played a leading role in the Swatch Group's expansion during the second half of the 1980s, as Italy was one of the company's top markets.

This new generation of managers brought with them new marketing concepts, which were instrumental in renewing brand image. One of the major innovations borrowed from the fashion industry was the concept of *flagship stores*, sales outlets which market a single brand, "where the client can live the brand".[18] They feature a uniform design and atmosphere throughout the world and are generally located on the main shopping streets of the planet's major metropolises, like the Champs-Élysées in Paris, Fifth Avenue in New York, and Ginza in Tokyo. In the case of the Swatch Group, these flagship stores were first introduced with the Swatch brand, the first Megastore opening in 1996, in New York. The following year, the Swatch Group opened some 60 Swatch Stores all over the world, doubling the number of outlets selling the brand in one fell swoop. According to Morrison and Bouquet, in 1998, the Swatch Time Shop in Times Square, New York, sold nearly 100,000 watches, which works out to a daily average of more than 270 pieces.[19] Finally, in the year 2000, capitalizing on the anti-conformism associated with the brand and its worldwide success as a watchmaking product, Swatch inaugurated a flagship store in the select Place Vendôme in Paris, surrounded by the world's leading luxury jewellery brands.

As we can see, the Swatch had a major impact in terms of marketing strategy. The experiences with the brand during the years 1983–1990 were then adopted by the company's other brands. However, the Swatch Group's restored ability to compete was not marketing-related. It was based first and foremost on a sweeping transformation of production systems, which boosted product profitability.

4 Rationalization and Globalization of the Production System (1985–1998)

The rationalization of means of production within the Swiss watch industry, which was pursued in the years 1985–1998, was clearly a new phenomenon in this industrial branch, given the scope of the restructuring. Indeed, the merger of ASUAG and SSIH led to a conglomerate of weakly integrated watchmaking manufacturers, and the restructuring of the production of watches, movements, and watch parts posed a major challenge in the first years of the Swatch Group's existence.

Of course, as we have seen, from the late 1960s onwards, many holding companies had been set up, bringing together several watchmaking firms, such as Chronos Holding (1966), the Société des Garde-Temps (1968), Saphir (1969), and the Société Anonyme des Fabricants Suisses d'Horlogerie (SAFSH, 1970). However, this initial phase of concentration was primarily aimed at streamlining the distribution system between complementary brands.[1] It did not involve genuine industrial restructuring, i.e. the concentration of production facilities. Likewise, the streamlining measures implemented before 1985 by various companies, such as Omega in Biel and Tissot in Le Locle during the interwar period, and their impact on overall industrial organization, were very limited.[2] As a result, the rationalization introduced by Nicolas Hayek appeared as a novel step, insofar as it marked the first time that the Swiss watch industry had opted for genuine, large-scale concentration of production structures, implying the closure of factories and the relocation of workers. Hayek Engineering's nearly 25 years of consulting experience with industrial firms from other sectors, in particular the German steel and metalworking industry (AEG, Krupp, Mercedes, Volkswagen), where the rationalization of means of production was a crucial issue, no doubt played a decisive part in the adoption of similar restructuring models by the watchmaking industry.[3]

Moreover, the rationalization of means of production went hand in hand with the globalization of production systems, characterized by the successful opening of production centres elsewhere in the world. This new industrial policy, which was conceived by Hayek, was adopted first by Ernst Thomke, Director General of the watchmaking division, then by Anton Bally from 1991 onwards.

Rationalization of production in Switzerland

One of the first steps taken was to centralize the production of watch movements and parts within ETA SA. This policy was not new as such but marked the continuation and culmination of the strategy for merging subsidiaries of Ebauches SA – the former name of ETA SA – which Thomke had implemented after he was chosen to head the company in 1978. The aim of this policy remained the same over time: the concentration of production was supposed to lead to rationalization (a reduction in the number of models, interchangeability of parts) and automation of production. However, the application of this policy to the Group as a whole led to major disruption of its main watch factories, which were obliged to stop manufacturing their own movements and focus on marketing.

Thus, rationalization resulted in functional dissociation through the Group, with some companies responsible for producing movements and others for selling watches. This role change significantly affected the Group's most prestigious watchmaking factories (Omega and Longines), which had hitherto manufactured internally the watch movements they then sold on the market but which were now obliged to source their movements from ETA SA. Accordingly, rationalization led to the relocation of employees between Swatch Group companies. For example, at Omega, the closure of the movement blank factories, a decision taken in 1985, led to the transfer of 190 of its 545 employees (or 34.9 per cent) to ETA SA the following year. A similar process could be seen at Longines, with the gradual transfer of production to ETA SA: supervision of production activities (1984), production of movement blanks (1986), and finally the assembly and encasing of movements (1988). As for Rado, it stopped assembling movements internally in 1987 and shut its Derendingen subsidiary in the canton of Solothurn.[4]

The second thrust of this rationalization strategy was the modernization and extension of equipment. The annual reports produced during this period do not provide details of the nature of divestments, depreciation, or fixed assets, as the balance sheets published put "land, buildings, machinery, and equipment" under a single heading, which is too general to give a clear picture of efforts to modernize equipment. However, the annual report published in 1990 mentioned that since 1986, the Swatch Group had invested over CHF1 billion in new production equipment, an amount that exceeds the total investment by SSIH and ASUAG individually during the 15 years prior to their merger (1968–1983). Starting in 1997, annual reports began to provide more details, making it possible to differentiate the category "plants and machinery" included under fixed assets. The amounts were CHF490 and 608 million in 1997 and 1998, respectively, representing a relatively high share of assets (13.5 per cent and 14.6 per cent). Fixed assets overall accounted for a stable share (an average of 31.1 per cent of assets in 1985–1992, followed by a slight drop to 24.3 per cent in 1998). Yet their share was never as high as the Group's equity capital, which shot up from 32.1 per cent of assets in 1985 to 60.2 per cent in 1992 and 71.5 per cent in 1998. Accordingly, the modernization and expansion of production machinery is to be seen as part of a financial strategy aimed at decreasing dependency on banks. The company's high profits, which amounted to a cumulative total of CHF1.9 billion for the period 1990–1995 (net income after tax), allowed the Swatch Group to more than double its reserves (CHF2.3 billion in 1995 as compared with CHF1.1 billion in 1990) and consider developing (equipment modernization, business acquisitions) without relying on banks or financial markets.

The expansion and modernization of production facilities also took the form of the acquisition of companies, primarily existing subcontractors of the Swatch Group. Apart from assembly plants in Asia and America, which will be covered below, mention should be made of the purchase of two Swiss firms, Georges Ruedin SA, a watchcase maker (1989) and Méco SA, a crown maker (1990). In France, the Swatch Group acquired two parts makers, Régis Mainier SA (1987) and the Marc Vuilleumier workshop (1990), which were subsequently merged into the Fabrique de fournitures de Bonnétage SA (1991), as well as the company Frésard Composants SA (1991). In Germany, the Swatch Group snapped up a movement and parts manufacturer, Pforzheimer Uhrenwerke PORTA GmbH (1990), which was renamed ETA Uhrenwerke shortly thereafter (1993). Thus, buying up production firms in Europe was part of a vertical integration strategy, which enabled the Swatch Group to tighten its grip on the sourcing of parts.

The globalization of production systems

The international division of production appears as the second main thrust of the strategy for streamlining production adopted by the Swatch Group in the 1980s and the 1990s. This was not, strictly speaking, a new strategy in the Swiss watch industry. The liberalization of the *Statut horloger* (Watchmaking Statute) in the course of the 1960s paved the way for foreign direct investment, and the main watchmaking groups opened production units abroad during the 1970s, especially in South-East Asia. For example, back in 1979, ASUAG had an interest in the company BETA SA, in Brazil, which reassembled watch movements and assembled watches. As for SSIH, it owned the Hong Kong-based firm Economic Swiss Time, which specialized in the production of pin-lever watches. Thus, the strategy for internationalizing production which the Swatch Group adopted from the outset was to a certain extent based on an existing legacy, but one which was rationalized and developed on a larger scale.

Opening a production centre in Asia was a management priority as soon as the rationalization push began. In 1985, ETA SA opened a factory in China, which was moved to Thailand the following year due to administrative difficulties with the Chinese bureaucracy. The company ETA (Thailand) Co. Ltd, based in Bangkok, specialized in producing watch components – primarily intended for Swiss plants – and assembling electronic modules for the Asian market. In 1989, it broadened its activities to include the production of watchcases when it bought up Leader, a majority Swiss-owned firm founded in

Bangkok in the early 1980s.[5] ETA's rapid development in Thailand was also marked by the opening of a branch in Malaysia, Micromechanics Sdn Bhd (1991). In 1998, Thai exports of watch movements, watch parts, and clocks totalled a hefty USD137.4 million, broken down as follows: USD45.8 million for Switzerland and 72.2 million for Hong Kong.[6] Expansion in Asia continued with the opening of a production centre in China, the company Zhuhai SMH Watchmaking Ltd, in the Shenzhen Special Economic Zone (1996).

The initial purpose of this Chinese subsidiary was to produce watch components – most likely for watches assembled in Switzerland – as well as quartz watch movements for world markets, primarily the United States, through Hong Kong-based firms. Moreover, the Swatch Group took advantage of its opening of a factory in China to try to launch Lanco, a watch brand for entry-level products. As Nicolas Hayek explained in a 1997 interview with the Geneva daily *Journal de Genève*, the Lanco product offering was a line of inexpensive watches, sold for a few score francs and produced in the Group's Chinese factories for world markets, especially the US market, where at the time nearly 80 per cent of all watch imports were in the USD20–25 price bracket.[7]

This decision was also the outcome of the development of the firm Shenzhen Fiyta Holdings.[8] Founded in 1987, this company was a spin-off of the watchmaking division of the industrial group belonging to the Chinese Air Force (China Aviation Technology Import and Export Company), which at the time had eight factories in the Shenzhen area. These were fitted out with machine tools and equipment imported from Switzerland and manufactured high-quality watches, featuring quartz movements imported from Japan and Switzerland with a design similar to some of the best-known Swiss models (Omega, Longines, Rado and TAG Heuer).[9] However, the Swatch Group did not intend to abandon the Chinese market for inexpensive watches to domestic rivals, which was what led Hayek's company to set up in China and launch the Lanco brand there.

Yet designing Lanco watches for the Chinese public was a side show which cannot explain on its own the Swatch Group's decision to invest in the country. In the 1990s, the main reason why American, European, and Japanese industrial companies opened production units in China was to take advantage of low-cost local labour for products intended for world markets, not to penetrate the domestic

Chinese market. Watchmaking was no exception, and China also became the world's watch factory.

The other event which no doubt helped convince the Swatch Group to open a subsidiary in China was the arrival of Citizen Watch Co. in 1994, with an initial production capacity of a million movements.[10] During the 1990s, this Japanese watchmaking group adopted a strategy of converting to the production of analogue quartz movements rather than complete watches. As a result, its production volume for watches and watch movements skyrocketed, rising from 65.8 million pieces in 1985 to 146.2 million in 1990, then soaring to 317.6 million in 2000.[11] No breakdown by product is available, but the extremely rapid rise in production volume says a great deal about this strategy of specializing in movements. By mass-producing electronic movements in its various factories in Japan and abroad, including China, Citizen aimed to dominate world markets in this field by offering very low-cost movements. For Switzerland, which had not yet repositioned to the luxury sector and for which quartz watches represented more than 90 per cent of all watch exports in 1990, this posed a real threat, especially in view of the fact that the share of movements within Swiss watch exports remained high (28.3 per cent of the volume of exports in 1990). Thus, Zhuhai SMH Watchmaking Ltd helped ensure that the Swatch Group remained competitive on world markets for quartz movements.

What is more, the Chinese subsidiary Zhuhai SMH Watchmaking Ltd supplied the Swiss factories of ETA SA with spare parts, the exact nature of which is not known. The "Swiss Made" legislation authorizes the use of parts produced abroad for the production of watch movements up to 50 per cent of the value of the parts as a whole.[12] However, it is extremely difficult to arrive at a clear picture of this international division of labour, as Swiss and Chinese foreign trade statistics yield completely different profiles.[13] According to the Swiss figures, between 1996 and 2005, Switzerland imported an annual average of USD2.8 million worth of parts and assembled movements from China, which represents a negligible business volume. Yet Chinese foreign trade statistics quote completely different figures: during the same period, China reportedly exported on average USD31.1 million of spare parts and movements. However, as the Chinese statistics show a sharp drop in exports to Switzerland in 2006 (USD5.1 million), precisely the year in which the Swatch

Group pulled out of China and relocated its Asian parts production to Thailand, the Chinese figures are probably closer to reality. As the bulk of the supplies were quartz movement parts, it is possible that they were declared as "electronic components" to the Swiss customs. For example, the subsidiary Zhuhai SMH Watchmaking Ltd reportedly supplied the Swiss plants of ETA SA with cheap spare parts, helping ensure that Swiss watches could compete on cost.

The economic directory *China Markets Yearbook*, published on a yearly basis in Beijing, lets us evaluate the relative importance of the Swatch Group factory in Shenzhen. According to this source, turnover for this firm showed steady growth, increasing from USD39 million in 2001 to 62 million in 2005. These figures provide a plausible account of the trade in parts with Switzerland, as captured by Chinese foreign trade statistics. For the Swatch Group, however, this was a minor subsidiary, as its total sales came to around 1 per cent of the Group's consolidated turnover. Nevertheless, despite its low financial significance, the Shenzhen factory had considerable strategic importance. It allowed the Swiss group to be present on the market for cheap electronic movements and to prevent rival companies from expanding in this segment, then moving upmarket. In 2005, the Swatch Group was well out in front in the ranking of the ten largest watchmaking companies in China (see Table 4.1). Zhuhai SMH Watchmaking Ltd boasted a much larger turnover than all its rivals, most of which were watchmaking firms belonging to overseas Chinese (Hong Kong, Macao) or the State, as the two companies specializing in the production of watches and movements, Longhai Huayi and Tianjin, were positioned at the low end intended for the domestic market and thus did not compete with the Swatch Group. Clearly, its main rival was the Japanese firm Citizen Watch Co., which, even though it was ranked only tenth according to this source, had a much larger market share. Its apparent weakness was due to the fact that only the group's Chinese headquarters in Beijing was registered, but it had a vast manufacturing network on Chinese soil, allowing it to produce several hundred million movements a year.

The Swatch Group had a second movement assembly centre outside Switzerland, in the Americas, with two main subsidiaries. The first was an assembly plant opened in 1961 by American investors in the Virgin Islands, Unitime Co. Ltd.[14] This company belonged to the

Table 4.1 The main watchmaking companies in China by turnover, in millions of yuan (RMB), 2005

Name	Ownership	Year founded	City	Province	Turnover (millions of yuan)	Number of employees	Main products
Zhuhai New Pearl Watchmaking Co. Ltd	Foreign company (Swatch Group, Swiss)	1996	Zhuhai	Guangdong	546.7	1,627	Movements and watch parts
Yantai Polaris Clock and Watch Group Company	State	1987	Yantai	Shandong	204.7	2,692	Clocks
Fuzhou Reida Electronics Co. Ltd	Overseas Chinese (Hong Kong)	1993	Fuzhou	Fujian	180	375	Quartz clocks
Zhaoqing Liheng Plastic Cement Hardware Electron Factory	Public enterprise	1992	Zhaoqing	Guangdong	163	480	Electronic watches
Glama Handicrafts (Shenzhen) Co. Ltd	Overseas Chinese	1990	Shenzhen	Guangdong	154.4	3,310	Clocks
Panyu Pearl Clock Factory	Overseas Chinese	1996	Guangzhou	Guangdong	151.7	521	Quartz clocks
Longhai Huayi Clocks and Watches Co. Ltd	Private entreprise (non-listed)	1993	Zhangzhou	Fujian	144.4	900	Watch movements
Tianjin Bohai Zhongou Watch Industry	Private entreprise (non-listed)	2002	Tianjin	Tianjin	133.6	1,053	Watch straps
Bode Watch Co. Ltd	Overseas Chinese	1994	Shenzhen	Guangdong	126.1	850	Clocks
Citizen (China) Clock & Watch Co. Ltd	Foreign company (Citizen Watch Co., Japan)	1994	Beijing	Beijing	122.9	108	Watch straps

Note: 100 yuan = 11.4 US dollars (between 1993 and 2006).

Source: *China Markets Yearbook*: 851 and *Chugoku shinshutsu*.

American watchmaking group Harris, which was bought out in 1973 by the firm Mido, G. Schaeren & Cie SA, a member of the holding company General Watch Co., itself controlled by ASUAG, with a view to expanding its activities on the US market.[15] The second assembly plant in the Americas was also a pre-merger relic: the company SMH does Amazonas, in Brazil. Even though the firm appeared only in 1990 as an operational company in the Swatch Group's annual reports, it was clearly the successor to BETA SA, in which ASUAG had had a minority stake.

Employment trends for the Swatch Group clearly reflect this globalization of production systems: the share of employees living in Switzerland, which stood at 80 per cent in 1983–1985, fell to 71 per cent in 1990, then to 54 per cent in 1998, remaining stable at this level up to 2010. This internationalization was primarily driven by the development of factories in the Far East. The share of Swatch Group employees living in Asia was 21 per cent in 1992 and 33 per cent in 1998. These figures include not only workers in the Swatch Group's Asian factories but also employees of sales companies and after-sales service networks. However, in view of the fact that sales companies in Asia did not really take off until after 2000, it is no exaggeration to say that more than one-fourth of all Swatch Group employees were Asian workers in the second half of the 1990s. ETA (Thailand) Co. Ltd alone employed nearly 3,000 people in 1994, that is, some 18 per cent of all Swatch Group employees.

Thus, in the decade following its founding, the Swatch Group restructured its production systems and deployed them internationally, primarily in Asia. However, the data needed to understand how this system worked are very fragmented. ETA does not provide any information on its foreign subsidiaries and has even gone so far as to conceal their existence. The company's Internet site shows only production sites in Switzerland on its interactive map of the different production sites.[16] Thus, on account of the lack of detailed data in the Swatch Group's annual reports on the activities of the various subsidiaries, it is not possible to identify the supply channels for parts or the structure of its outsourcing network. However, Swiss foreign trade statistics are a valuable tool for shedding light on some key elements of the global value chain in the watch industry.

The global value chain in the watch industry in 1995

In 1995, the main producers of movements in Switzerland were the Swatch Group (more than 100 million movements, quartz and mechanical), ISA Swiss SA (some 45 million movements, quartz), and Ronda SA (some 10 million movements, quartz). There were also some small manufacturers of mechanical movements, mainly in the luxury segment.[17] Thus, the Swatch Group had a market share of around two-thirds for quartz movements and more than 90 per cent for mechanical movements. In 1995, the two other quartz movement producers also had production centres in Asia: ISA had subsidiaries in Hong Kong and in China (ISA Far East),[18] while Ronda had a sales subsidiary in Hong Kong from 1970 and a production centre in Thailand from 1972.[19] At the time, these three companies were positioned on two different markets: the sale of "Swiss Made" movements to Swiss watch assemblers, and the supply of movements made in Asia but without the "Swiss Made" label for the global market.

Using the Global Trade Atlas database makes it possible to access data on foreign trade in watch parts between the main countries in the world and to visualize the related global flows (see Figure 4.1). On account of differences between the data published by different countries, the values used here are the average of the value of exports from Country A to Country B (data from Country A) and the value of imports to Country B from Country A (data from Country B). Global flows of watch movement parts shed light on the existence of a production system which was already deeply globalized in the mid-1990s, as the main watchmaking nations were interdependent. Three features can be underlined.

First, Switzerland's reliance on foreign nations for movement parts was low: it exported significantly more (USD256.7 million) than it imported (USD171.3 million) in 1996. Imports came primarily from Thailand, which hosted subsidiaries of the Swatch Group and Ronda, and from Hong Kong – which re-exported parts made in China to Switzerland. As for exports of parts, they were mostly intended for Thailand, but also for Hong Kong and China.

Second, Hong Kong was a veritable hub as far as trade in watch parts and movements in Asia was concerned. It played a key role with respect to the assembly of cheap quartz watches for the global market. Thus, in 1997, except for parts, which were mainly provided by Japan

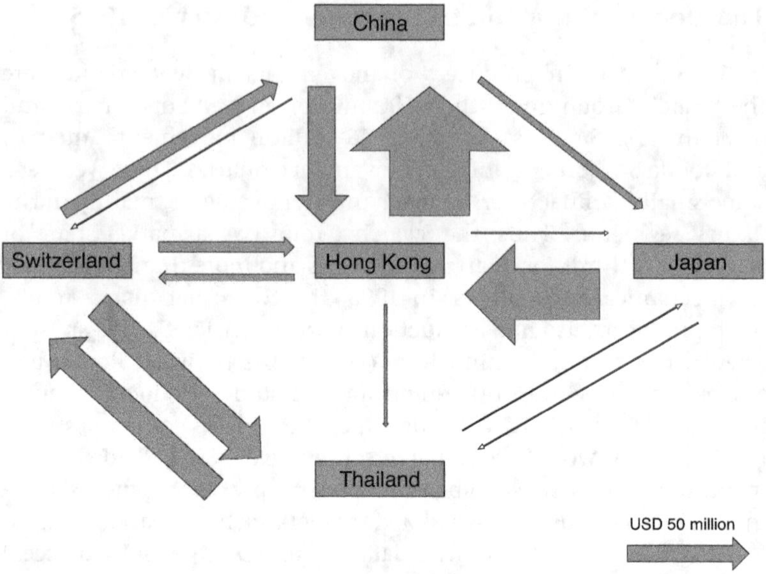

Figure 4.1 Global flow of watch movement parts, value in USD, around 1995

Note: To ensure legibility of figures, flows of less than USD5 million are not shown. Flows between Switzerland and Thailand are based on 1996 data (1995 not available in *World Trade Atlas*), while data for Hong Kong are based on 1997.

Source: *Global Trade Atlas*, http://www.gtis.com, category 9114 (watch parts).

and China, Hong Kong supplied a large amount of movements from Japan (USD759.7 million), China (USD187.4 million) and Thailand (USD81.5 million), but also cases from China (USD52.8 million) and Japan (USD21.5 million). The watches assembled in Hong Kong were mainly exported to the United States: in 1997, their overall value amounted to USD3.8 billion, of which 29.4 per cent went to the American market and 9.7 per cent to Japan, the second largest outlet.

Third, China had not yet become a major production centre for the world watch industry. In addition to parts, it imported movements, mainly from Japan (USD406.6 million) and Hong Kong (USD124.2 million). These were assembled with cases made in China or supplied by these two countries, then exported, to the tune of USD1 billion in 1995, to Hong Kong (31.2 per cent, probably re-exported afterwards), Japan (28.8 per cent), and the United States (21.6 per cent). At the time, each nation had a specific use for China: whereas the watch traders of Hong Kong owned assembly and production plants for supplying cheap quartz watches to the American market, the Japanese companies relocated part of their production for their domestic market there. As a result, a flow of parts between China and Japan, via Hong Kong, can be observed. As for Switzerland, it was weakly integrated into the Chinese production network for watches until the Swatch Group opened a subsidiary in Shenzhen in 1996.

Thus, when not only the Swatch Group but also its rivals Ronda and ISA opened production subsidiaries in Asia, their aim was not primarily to set up a global production network for their own needs. Of course, Thailand was a major centre for the production of cheap parts for the Swiss watch industry. Yet this industry is definitely less globalized than the competitors from Japan and Hong Kong, whose production systems are rooted within a global organization. Trade statistics for watch parts highlight commercial flows from Japan to Hong Kong then China, where the assembly centres for the main manufacturers of Japanese watches are located.

Production geared to watch movements

The Swatch Group's annual reports contain very little information on the type of products manufactured within the framework of this rationalized, globalized production system. Information by product is virtually non-existent, with the exception of production statistics published in 1986–1987 and 1992–1998. An analysis of these data sheds light on the impact of rationalization in terms of product development.

Swatch Group production in 1986–1987

Data published in its 1986 and 1987 annual reports show that the Swatch Group produced essentially movements (see Table 4.2): their volume amounted to more than 30 million pieces per year, as against 15 million for complete watches. Moreover, these were mostly quartz movements (98.4 per cent of volume and 95.3 per cent of value). The high proportion of movements can be explained by the fact that one of the two enterprises merged into the Swatch Group in 1983,

Table 4.2 Product sales of the Swatch Group, thousands of pieces, 1986–1987

	1986				1987			
	Movements		**Watches**		**Movements**		**Watches**	
	N	%	N	%	N	%	N	%
Switzerland	11,349	36.9	1,400	9.4	10,040	32.8	1,420	8.6
Europe	1,494	4.9	6,034	40.3	846	2.8	6,762	40.7
Far East	10,153	33.0	463	3.1	12,510	40.9	762	4.6
America	7,484	24.3	6,169	41.2	7,192	23.5	6,700	40.4
Other	308	1.0	893	6.0	3	0.0	957	5.8
Total	**30,788**	**100**	**14,959**	**100**	**30,591**	**100**	**16,601**	**100**

Source: Swatch Group, *Annual Report*, 1987: 10.

ASUAG, was actually a trust that had controlled parts and movement blank production for the entire Swiss watch industry since the 1930s.

Accordingly, the Swatch Group perpetuated its historical role as movement supplier for virtually all Swiss watch enterprises. Thus, the proportion of movements sold in Switzerland came to 36.9 per cent and 32.8 per cent of the total, for volumes in excess of 10 million pieces. These are very large volumes, which reflect the Swatch Group's domination of the industry. If we accept that around 90 per cent of these movements were then exported as complete watches by their buyers – that is, about the proportion of exports by the Swatch Group – this means that watches equipped with Swatch Group movements represented 70.4 per cent of all Swiss watches exported by independent makers in 1986 and 72.5 per cent in 1987.[20] Thus, the Swiss watch industry as a whole relied heavily on the Swatch Group for parts and movements when growth returned in the mid-1980s.

As was only natural, movements accounted for substantially lower average value (CHF10.6 in 1986) than watches (CHF83.1 in 1986). Yet the sale of movements was important for the Swatch Group. In 1986, even if it totalled a mere CHF326 million, that is 17.9 per cent of gross sales, this was an amount similar to that for exports of non-metallic watches (Swatch) that same year. The new-found growth which the Swiss watch industry experienced from 1985 onwards, with a volume of completed watches rising from less than 20 million pieces per year in 1982–1984 to 21.1 million in 1985 and 34.9 million on average in the 1990s, undeniably benefited the main movement producer. In 1988, however, the Swatch Group stopped providing information on movement volume and shares in its annual reports. The firm's communication strategy thereafter emphasized the worldwide success of the Swatch as the key profit source, and began to hide the fact that it was first and foremost a movement producer. Nevertheless, in 1997, the value of the division "movements and components" represented sales of CHF1.2 billion, or 32.2 per cent of gross sales. Thus, this division apparently grew steadily over a ten-year period.

Finally, the point should be made that the Swatch Group did not sell watch movements only to Swiss companies. Its main customers for this product in 1986 and 1987 were actually foreign companies, primarily Asian and American ones. It is not possible to know the nature of these movements, but they apparently involved products made in the Swatch Group's plants abroad.

As for complete watches, the only data published are those concerning the years 1986 and 1987. Two comments are in order. First, the domestic market appears to be more important than is often thought, as it amounted to some 10 per cent of volume. Second, the figures underscore the Swatch Group's heavy reliance on the large, traditional outlets: Europe and the United States. The Swatch Group, which exported 90 per cent of its finished watches to these two destinations, was even more dependent on them than the Swiss watch industry as a whole (83.4 per cent of exports).[21] While Europe was still a competitive market for the Swiss watch industry, due to tight control of distribution networks and a competitive advantage of Swiss brands over American and Asian ones, the Unites States posed a problem for the Swatch Group. The heavy reliance on this market, which represented more than 40 per cent of sales volume in 1986 and 1987, primarily reflects the absence of other fast-growing markets, especially in Asia. The share of the Far East came to a paltry 3.1 per cent of sales volume in 1986 and a meagre 4.6 per cent in 1987. For the entire Swiss watch industry, the three biggest Asian outlets – Hong Kong, Japan, and Singapore – together accounted for a mere 5.7 per cent of exports of complete watches in 1986.

No figures for complete watches have been published since 1987. However, there is information on the geographical breakdown of sales for the watchmaking division for the years 1986 and 1992

Table 4.3 Geographical breakdown of sales of complete watches and movements by the Swatch Group, as a percentage, 1986 and 1992

	Volume		Value	
	1986	1992	1986	1992
Europe	44	49	49	65
America	30	8	29	13
Asia	24	49	20	21
Other	2	1	2	1

Source: Swatch Group, *Annual Reports*, 1986 and 1992.

(see Table 4.3), making it possible to determine the overall trend for Swatch Group markets during this period. First of all, there was an upturn in the European market, as reflected not only by an increase in sales of movements and movement blanks to other Swiss watch manufacturers but also by a sharp rise in value accompanied by a quasi-stagnation of volume, pointing to a significant increase in the sale of finished products in Europe. Next, there was a major shift from America to Asia. The New World was in free fall, in terms of both volume (from 30 per cent to 8 per cent) and value (from 29 per cent to 13 per cent), no doubt reflecting a drop in sales of both watch movements and finished watches. On the American market, Swiss watchmakers found themselves up against inexpensive quartz watches imported from Hong Kong, a trend which was reinforced by the opening of a Citizen Watch Co. firm in China (1994), making it possible for the Swatch Group's rivals to source quality movements at rock-bottom prices. Asia saw its consumption of watch movements soar, reflecting the growth of ETA (Thailand) Co. Ltd: although its share of sales in terms of value did not budge, its share of volume doubled, as a result of which Asia was absorbing half of the Swatch Group's total production volume (watches and watch movements) at the time.

Developing business by growing volume (1992–1998)

Sales statistics for the years 1992–1998 help us understand changes in product types during the 1990s (see Table 4.4). The main characteristic is a drop in the average value of pieces, from CHF28 in 1992 to CHF22 in 1998. This fall is no doubt indicative of structural trends

Table 4.4 Watch and movement production by the Swatch Group, 1992–1998

	1992	**1993**	**1994**	**1995**	**1996**	**1997**	**1998**
Volume, millions of pieces	86.9	86.8	94.8	101.4	102.5	116.1	118.8
Value, CHF millions	2423	2430	2228	2184	2266	2487	2630
Average value, CHF	28	28	24	22	22	21	22
Asia, as a % of value	21	25	28	29	28	29	27
America, as a % of value	13	12	11	11	13	13	15
Europe, as a % of value	65	61	58	58	57	56	57
Asia, as a % of volume	49	49	61	62	61	70	68
America, as a % of volume	8	7	3	3	5	4	4
Europe, as a % of volume	42	43	35	34	33	26	28

Source: Swatch Group, *Annual Reports*, 1992–1998.

in production, marked by an increase in the share of movements, as they are much less expensive than finished watches. Moreover, Asia's growing importance as an outlet tends to confirm this hypothesis, as this world region was primarily a consumer of watch movements at the time. Indeed, Hong Kong watchmakers shifted in the 1990s to the production of quartz watches under licence for fashion industry clients or for non-watchmaking sectors all over the world. Having transferred their manufacturing activities to neighbouring China, they became part of global value chains and acted as middlemen between cheap Chinese labour and global markets. They sourced movements from Swiss and Japanese firms, which they proceeded to insert into watchcases manufactured in China, then exported finished watches to clients all over the world, in particular in the United States.[22]

Clearly, the Swatch Group moved into the market for supplying Hong Kong watchmakers with quartz movements during the 1990s: Asia's share of the volume of company sales (finished watches plus movements) recorded a steep increase, rising from 49 per cent to 68 per cent of the total between 1992 and 1998. Moreover, the drop in the average value of pieces sold in Asia, which plummeted from CHF15.90 in 1992 to CHF7.20 in 1998, confirms the tendency towards mass purchases of movements in this part of the world. This involved pieces which were primarily manufactured in the Group's Asian factories. Indeed, ETA (Thailand) Co. Ltd experienced a remarkable boom during the 1990s.

Europe – Switzerland included – maintained its key role but showed some changes. It was still the main source of gross sales, given that the growth of movement sales in Asia had a weak impact on the relative weight of Europe (65 per cent in 1992 and 57 per cent in 1998). As for Europe's relative decline in terms of volume, it can be explained by the general high growth in sales volume. In absolute numbers, the European market was stagnating rather than dropping, with an average volume of 34.1 million pieces a year. In 1986–1987, the European market absorbed only some 20 million pieces a year, of which more than half were movements for other Swiss watch companies. As a result, the growth of the European market between 1986/1987 and the 1990s can be explained by the sale of complete watches, on the one hand, but also by trade with rival Swiss companies, a major outlet, on the other hand. Indeed, the overall volume of Swiss watch exports went from 27.6 million pieces in 1987 to an average of 30.6 million in 1988–1991, then grew again to an average of 38.2 million in 1992–1996 before dropping to under 33 million in 1997–1998. There was thus a general expansionary trend until 1996 and the Swatch Group was the main supplier to the industry, as can be seen from the decline in the average price of pieces sold in Europe between 1992 (CHF43) and 1993 (CHF38). Moreover, the drop in export volume in 1997 probably led to a decrease in orders for the Swatch Group, as reflected by a new increase in the average price (CHF46 in 1997; CHF45 in 1998). Finally, since 1997, the Swatch Group has published more detailed data on the composition of its gross sales. They confirm the strategic importance of the sale of movements to rival Swiss companies. In 1997, completed watches amounted to 59.3 per cent of gross sales, as compared with 32.2 per cent for the "movements and components" division.

Attempts to diversify into new technologies

The involvement of the Swatch Group in businesses other than watchmaking dates back to the Group's foundation and stems from the production of electronic components for quartz watch movements. At the time of its creation, the Swatch Group inherited several firms belonging to ASUAG and engaged in R&D and production of electronic technologies for watchmaking. The main companies were EM Microelectronic-Marin SA (chips), Oscilloquartz SA (quartz crystals), Renata SA (batteries), Lasag SA (industrial lasers), and Asulab SA (R&D). Its command of such technologies in electronics led the Swatch Group to consider diversifying into new sectors.

The importance which Nicolas Hayek attached to high technologies can, however, be explained by his determination to stay at the forefront of the field of watchmaking. Since 1993, these companies have had a representative on the Swatch Group's Extended Management Board, in the person of the director of EM Microelectronic-Marin, Mougahed Darwish, and have been the object of steady investment, notably through cooperation with external partners. For example, agreements were signed with the Swiss Federal Institute of Technology Lausanne and the French industrial group Thomson-CFS for R&D in lasers (1986) and with the Dutch multinational Philips for the production of chips in Switzerland (1987). In parallel, the Swatch Group has rationalized its high-tech companies by selling off non-strategic sectors. For example, Lasag SA span off its ophthalmologic laser division in 1992. (Subsequently, it was bought by the American multinational Rofin-Sinar Technologies Inc. in 2010.)

The main feature of the 1990s was in fact diversification into two new high-tech sectors: automobiles and telecommunications. In the case of automobiles, the Swatch Group committed to manufacturing an ecological car. After collaborating with the Biel School of Engineering (1990), the Swatch Group entered into an alliance with Volkswagen to launch the manufacturing process (1991), but differences of opinion between the two partners led to the severance of ties in 1993. Hayek then turned to Mercedes-Benz with the Smart project (1994). Two joint ventures were established in 1996: Micro Compact CA SA, in Switzerland, with a capital of CHF300 million (49 per cent Swatch Group), in charge of R&D, and MCC Micro Car SA, in France, with a capital of FF100 million (37 per cent Swatch Group), for production. The Smart car was born from this cooperation and was launched in November 1997, but without a hybrid motor for technical reasons. As for the second sector, telecommunications, it was supported by the subsidiary Swatch Telecom SA (1996). This company launched telephones with a marketing strategy inspired by Swatch, that is, fashionable and cheap products.

However, despite its efforts to diversify, the Swatch Group remained essentially a watch group. The comparison between 1985/1986 and 1997/1998 shows that the "technology" division posted only slow growth: it amounted to CHF279 million per year in 1985/1986 and 347 million in 1997/1998. In terms of percentage of gross sales, this division was even declining, with a drop from 15.1 per cent to 9.2 per cent.

At the end of the 1990s, the Swatch Group decided to disinvest from technologies unrelated to watches and to focus on core competences. Despite the fact that Nicolas Hayek reaffirmed in 1999 his desire to "open the spectrum of watchmaking",[23] the technology division did not record any special growth. Even if sales went from CHF393 million in 1998 to a peak of 630 million in 2007 before dropping to 380 million in 2009, the division's relative share stagnated (9.8 per cent of gross sales in 1998–2009). Most of the diversifications outside watchmaking implemented in the 1990s were abandoned.

The main divestment concerned the automobile industry, with the withdrawal from the Smart project. In 1997 and 1998, the Swatch Group withdrew gradually from the company Micro Compact Car SA and sold its stake to Daimler-Benz. Also, the German company SID Sokymat Identifikations-Komponenten GmbH, which produced anti-skid systems for automobiles and whose sole chip supplier was EM Marin, was taken over by the Dutch company Smatrac N.V. (2008) a few years after being purchased by the Swatch Group (2003). In 2009, the company Microcomponents SA was sold to the Singapore-based group Juken Technology, which also took over some activities of the factory in Zhuhai (China) in 2010.[24] The same trend could be seen in the telecommunications sector, which was abandoned when Swatch Telecom was wound up in 2005.

Finally, the strategy of concentrating on core competences was implemented by several of the Swatch Group's parts manufacturers, which closed down divisions unrelated to watchmaking. For example, Nivarox-FAR sold its metrology division (gauges, measuring instruments) to Brown & Sharpe TESA SA in 2000. In addition, when Omega Electronics AG was liquidated in 2005, the two divisions linked to the watch business were taken over by two other Group companies, Swiss Timing (sports timing) and EM Microelectronic-Marin (radio frequency identification), while the division Passenger Information Systems was sold to PEL Passenger Electronics SA.

Moreover, the watch divisions of the company Michel SA were taken over by two other Swatch Group companies (ETA SA and Meco SA), before this firm was sold on to Ferton Holding SA, a producer of parts for the automobile industry (2008).

In parallel to these divestments, however, the Swatch Group strengthened the high-tech sectors linked to the watch business. The chipmaker EM-Microelectronic opened a subsidiary in the United States, EM Design Inc. (1997), and bought up the Czech company Asicentrum, the former microelectronics division of a State-owned R&D centre (2001). Moreover, the Swatch Group developed the sports timing sector with the acquisition of two firms in Germany and the Czech Republic (2006). This field, in which the Swatch Group is active through its subsidiary Swiss Timing, is essential for the watch trade in terms of marketing and communication. The Swatch Group competes with its Japanese rival Seiko to time the most famous sports events with a worldwide impact, such as the Olympics. Even though technical operations are carried out by Swiss Timing, depending on the event and the sport, the Swatch Group uses the brand best suited to the image conveyed (Omega and Swatch for the Olympics; Longines for gymnastics and horse riding; Rado for tennis; Tissot for basketball, cycling, ice hockey, etc.).

The Swatch Group was not the only firm to refocus on its core competences in the second half of the 1990s. Rather, this reflected a phenomenon which affected many multinationals in all of the world's industrial sectors from the 1980s onwards. In a famous article published in 1990 in the *Harvard Business Review*, Prahalad and Hamel stated that "managers have to win manufacturing leadership in core products and capture global share through brand-building programs aimed at exploiting economies of scope.... We believe an obsession with competence building will characterize the global winners of the 1990s."[25] In the watchmaking sector, the quartz revolution led watch manufacturing companies to master new technologies tied to electronics and to attempt to leverage this new know-how by diversifying into new activities. However, the Swiss watch industry, even after it had undergone restructuring and concentration into a few large companies such as the Swatch Group, soon reached the limits of its capacity for action in an extremely competitive electronics industry dominated by powerful American and Japanese multinational corporations.[26] Apart from a few specialized products and a handful of niche markets, like sports chronometry, Swiss watchmaking companies did not have the size or financial clout to compete in this new environment.

Consequently, the Swatch Group was not the only Swiss watchmaker to refocus on its core business, namely watchmaking. The same trend could be seen, at an earlier stage and on a different scale, with other watchmaking firms, which abandoned electronic technologies as early as the 1980s because they lacked the means to retain control of this industry. One of the best examples is no doubt the company Heuer-Léonidas SA, which went on to become TAG Heuer in 1985.[27] This family-owned company, which specialized in the production of chronographs and sports counters, opted in the 1970s to move into electronic devices, primarily sports counters. The electronics division, which was built around the US subsidiary Heuer Time & Electronics Co., was restructured in 1970 and experienced such strong growth that the company's managers announced in 1976 that their aim was "to scale the traditional watchmaking sector back to more modest dimensions".[28] Electronics' share of the consolidated turnover of the Heuer group shot up from 3.4 per cent in 1970 to 34.2 per cent in 1977. By 1980, wristwatches accounted for a scant 17.4 per cent of sales, compared with 70.8 per cent for sports counters. However, the arrival on the market of extremely cheap digital chronographs and sports counters produced in Hong Kong and Taiwan meant that Heuer-Léonidas was no longer competitive, and the company's attempts to diversify into electronic measuring equipment did not pan out. As a result, this small family firm, which at the height of its expansion in 1973 had a total of 338 employees working in its subsidiaries all over the world, was on the verge of bankruptcy. In 1982, it was bought up by the company Nouvelle Lémania SA (formerly Lugrin SA, see Chapter 3), a manufacturer of complicated watch movements which SSIH had acquired in 1932. Shortly thereafter, it was taken over by a group of investors including the luxury watchmaker Piaget (1981). In 1985, Heuer-Léonidas was sold on to the multinational enterprise Technique d'Avant-Garde SA (TAG), based in Luxembourg and specializing in technologies for maritime and air navigation, as well as Formula 1 car racing (it had an interest in the McLaren stable).[29] Heuer's ownership changes of the 1980s were accompanied by a strategic change in brand management and product development. The TAG Heuer company and brand focused on watchmaking and sports wristwatches, jettisoning its previous failed diversification into electronics.

However, the shift away from the strategy of diversification into electronics which could be seen in Swiss watch companies in the 1980s and 1990s was not characteristic of the industry as a whole. In Japan, what happened was precisely the opposite. The Seiko and Citizen watchmaking groups, which dominated the industry, began to diversify increasingly into electronic technologies in the 1980s, a trend which picked up steam as their watches became less competitive and is still under way today. In the case of the Seiko Holdings group, the two watchmaking factories which had made the company so successful from the 1950s onwards were restructured and repositioned in essentially non-watchmaking sectors. Daini Seikosha took the name of Seiko Instruments and Electronics (1983), then Seiko Instruments Inc. (SII) in 2004, moving into electronic devices and modules while continuing to produce watches and machine tools. As for Suwa Seikosha, it merged with its own subsidiary Epson, taking on the latter's name (1985) and going on to specialize in electronic printers and modules.[30] Following these changes, watchmaking's share of group turnover, which as recently as 1980 was still at 90 per cent, fell sharply to 48 per cent in 2000 and 33 per cent in 2010.[31] The Japanese example shows that the technological journey towards electronics was a possible development path for watchmaking enterprises. The issue was to have organizational capabilities to support this mutation, which the Swiss watchmakers lacked at that time.

The outcome: better productivity

As far as the Swatch Group was concerned, the policy of rationalization, refocusing on core competences, and globalization had two major impacts. First, it paved the way for the growth in production, which was still viewed as a key strategic goal up until the mid-1990s, as Swiss watchmakers still felt that turnover growth was a priority. The volume of watches and movements produced by the Swatch Group soared, increasing from 44.3 million pieces in 1985 to 86.9 million in 1992 and 118.8 million in 1998. The growth can be broken down into two distinct phases, with a sharp rise in 1985–1992 (when annual growth averaged 8.3 per cent), followed by a more sedate increase in 1992–1998 (5.3 per cent), a different pace due to changes in productivity. Second, it made the company competitive on world markets once again, by enabling it to contain production costs.

A comparison of trends for productivity and average value added per employee reveals two different phases in the company's development (see Figure 4.2). First of all, the period of 1985–1992, when the bulk of production rationalization took place, was marked by steep rises in both productivity and value added per employee. The average number of pieces produced per employee rose steadily, from 3,843 in 1985 to 6,075 in 1992. During this same period, value added by employee went from CHF53,444 in 1985 to CHF89,485 in 1992. As can be seen, Swatch Group's restructuring enabled its employees to produce more pieces and make more money for the company.

Second, the Swatch Group entered a new phase in the mid-1990s, when growth was no longer driven by productivity increases. The number of pieces produced per employee remained relatively stable, hovering around an average of 6,034 for the years 1992–2004. In 2005, the Swatch Group stopped publishing figures for production volume,

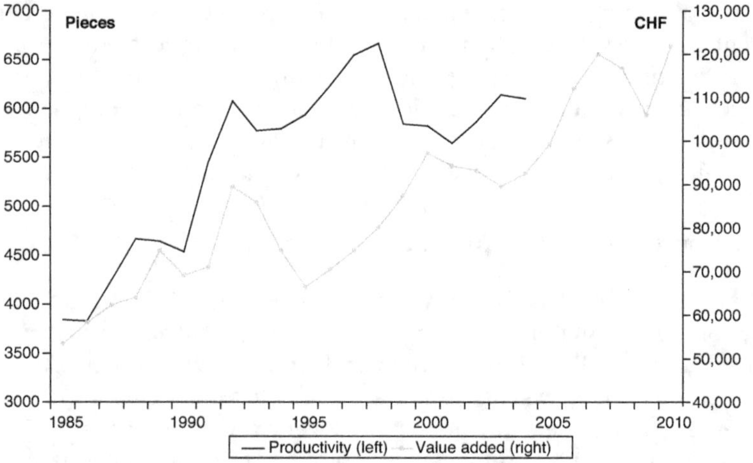

Figure 4.2 Productivity and net value added per employee, 1985–2010

Note 1: No figures given for production volume after 2005.

Note 2: Net value added = earnings after depreciation (before taxes and dividends) + payroll.

Source: Own calculations, based on Swatch Group, *Annual Reports*, 1985–2010.

but there are no signs pointing to an increase in volume since then. After the company concentrated production facilities in Switzerland and globalized its production system, it achieved optimum productivity. However, following a phase of decline and stagnation in 1992–1995, which was clearly due to the phenomenal growth of the Group's Asian factories, average value added per employee rose quite steadily, increasing from CHF66,444 in 1995 to CHF121,840 in 2010. This virtual doubling of value added per employee was primarily due to the Swatch Group's acquisition of luxury brands and the rise in growth of the Group's brands (Longines and Omega). For the past 15 years, the pursuit of growth has been based on the implementation of a new marketing policy (see below), not the rationalization of means of production and Fordian logic.

The financial impact of restructuring

The main financial consequence of restructuring the means of production and distribution systems was that the company gradually became more independent and no longer had to rely on bank loans. The Swatch Group became a very lucrative company, ploughing back a large share of its profits into developing the firm. Net profit after depreciation, reserves, and taxes posted continuous growth, rising from 3.3 per cent of gross sales in 1985 to 10.7 per cent in 1995. This healthy financial situation made it possible to distribute ever-larger dividends (5 per cent in 1986; 12 per cent in 1990; 18 per cent in 1992). In this way, the investors who supported Nicolas Hayek in 1985 were rewarded for their faithfulness.

Yet the relentless growth in profits should not obscure a key element of the firm's management: its self-financing policy. Modernization of the means of production and company growth went hand in hand with ever-diminishing reliance on banks. A huge amount of profit was ploughed back into reserves. Excluding paid-up capital, the Swatch Group's accumulated equity capital came to a whopping CHF2.0 billion in 1995, compared with a scant CHF190 million in 1985 and 750 million in 1990. When measured as a percentage of the balance sheet, shareholders' equity posted a rising trend until 1995. Whereas it was only 32.1 per cent in 1985, it passed the 50 per cent mark in 1989 and hit 70.1 per cent in 1995. This financial empowerment enabled the Swatch Group to develop without relying on bank loans. For Nicolas Hayek, who was only a minority shareholder in 1985, this strategy was the road to independence in the medium term. As for independence from the banks, their disinvestment from ASUAG and SSIH was an objective dating back to the 1970s. The development of the Swatch Group's governance clearly reflects this process of emancipation, with the successive departure of most of the bankers from the Board of Directors: Rolf Beeler (BPS) and Paul Risch (BCB) in 1993, followed by Walter G. Frehner (SBS) in 1997. Since 1998, there has been only one banker on the Board: Peter Gross, a UBS director.

Moreover, the Swatch Group's financial policy had another impact on governance, with the withdrawal of some investors from Hayek Pool, who apparently wanted a larger distribution of profits. This was above all the case with Stephen Schmidheiny, president of the investment company Anova Holding, administrator of many companies (UBS, Nestlé, ABB, Landis & Gyr) and a member of the Swatch Group's Board since 1986. He withdrew in December 1993 from the investment company Wat Holding, jointly owned with Nicolas Hayek, which was a major shareholder in Hayek Pool, and left the Swatch Group's Board in 1994. Some other investors did the same at the end of the contract binding the members of Hayek Pool (1997).

5
A New Marketing Strategy (1985–1995)

The restructuring and the globalization of the production system were not enough on their own to power Swiss watchmaking's triumphant return to world markets in the second half of the 1980s. Production was but a single aspect of a multifaceted policy, which also attached considerable importance to marketing strategy, whose role was to enable the Swatch Group to establish itself as the world's leading watchmaking enterprise. This new marketing strategy was implemented in two stages. First, during the years 1985–1995, the emphasis was placed on restructuring distribution systems, a move which was part of the overall policy to streamline the Group. Second, in the mid-1990s, the Swatch Group adopted a policy of brand repositioning.

Rationalizing distribution

From the mid-1980s onwards, the restructuring of production facilities was accompanied by an in-depth rationalization of distribution systems, characterized by the establishment of sales companies to service all group brands on the main markets while retaining certain agents. As a result, vertical integration was not complete, but varied according to the specific conditions in individual markets.

When the Swatch Group was founded, each watchmaking company had its own network of subsidiaries responsible for distributing its production, as a result of which the Swatch Group had several sales companies in certain countries in the middle of the 1980s. By way of example, in 1985, it had eight sales subsidiaries in the United States (Certina, Collector's Classic Edition Society, ETA, Hamilton Clock, Omega, Rado, Swatch, and Unitime), five in Germany (Certina, Endura, Longines, Omega, and Rado) and two in Hong Kong (Endura and Longines).

The principle of rationalization adopted in the second half of the 1980s led to the establishment of a single sales company per country for all group brands, a streamlining measure aimed at generating economies of scale, particularly through a reduction in fixed assets and simplification of logistics. Trends for the number of sales subsidiaries (20 in 1985, 19 in 1991, and 22 in 1998) do not really reflect this rationalization, because the latter went hand in hand with a geographical extension of the network of subsidiaries. On the one hand, the Swatch Group closed several subsidiaries. In 1991, it had only a single distribution company left in the United States. The same held true for Germany. Similarly, the Swatch Group wound up Longines SA for sales in Switzerland (1993), Omega Electronics Ltd in the UK (1994), SMH Trading Far East in Hong Kong (1994) and SMH Venezuela (1994). On the other hand, it opened a number of distribution subsidiaries in other countries: Brazil (1986), Ireland (1986), Malaysia (1988), Netherlands (1986, subsequently closed in 1987), Singapore (1986), Austria (1993), British Virgin Islands (1993), Canada (1993), Portugal (1993), South Korea (1994), and once again the Netherlands (1997).

One of the reasons for redeploying distribution networks was to rationalize logistics, as stocks of watches and parts were a major financial burden for the Group: in 1985, they represented 33.6 per cent of the balance sheet. Ten years later, this share had fallen to 25 per cent. However, the restructuring was also aimed at facilitating the adoption of a global marketing strategy. Indeed, the repositioning of brands requires direct control of individual markets, in order to ensure their new visibility and the dissemination of a standardized communication policy worldwide. Until the 1980s, however, local agents generally had considerable leeway in terms of image and publicity, or even watch design and pricing. This need to control – and standardize to a certain extent – product image is the main reason why this policy of geographical extension was pursued in the 1990s and 2000s. An analysis based on traditional marketing would consider this strategy to be a reflection of the globalization consumption of luxury goods, as characterized by a very high degree of standardization.[1] Yet, as Hanssens points out, watchmaking provides an illustration of an industry where products are only partially globalized and continue to be adapted to local cultural consumption patterns.[2]

Finally, since the end of the 1990s, the Swatch Group has focused increasingly on distribution activities. It has started to internalize retail sales, previously limited to jewellers and independent retailers. This new activity has allowed the Group not only to reinforce its control over brand image but also to internalize a greater share of turnover and profits.

Brand policy

Initially, centralization did not have any impact on the marketing strategies for the various brands. It primarily concerned logistics (administration, finance, after-sales service, and IT). In the second half of the 1980s, the Swatch Group as a whole did not have a brand policy as such, even though senior management decided to drop some underperforming brands which it had acquired during the 1983 merger targeted at the low end of the market, such as Atlantic Watch, Baume, Derby Watch, and Record Watch. This rationalization was designed to limit the number of brands, so as to improve positioning and boost complementarity between brands.

Likewise, the only two new brands launched during this period – Flik Flak children's watches (1987) and Pierre Balmain fashion watches (1987) – did not reflect a desire to diversify the Group's brand portfolio. Rather, this followed a completely different line of reasoning, which could be called an industrial logic. These two brands came straight from the Group's movement factory, ETA SA, in an effort to overcome the crisis and the keen competition from Japan.

Otherwise, differentiation between the brand images was not pronounced. Omega and Longines remained the historical rivals they had always been since the late 19th century, with various product lines embodying similar values (precision, fitness, fashion, elegance). There was also a certain amount of ambivalence in the middle of the range (Certina, Mido, Tissot), where differentiation was difficult, even though Tissot was chosen to launch new products like the Rock Watch, with a stone case (1985), the Wood Watch, with a wooden case (1988), and the Two Timer watch, with a dual analogue and digital display (1986). The only two brands which stood out from the crowd in terms of image were Rado, which relied on a modern design, and Hamilton – purchased by SSIH in 1971 – which cultivated its American roots (American Tradition line). This lack of differentiation is no doubt the reason it was thought necessary to start by reforming brand policy internally, within each company. The goal was rationalization through a drastic reduction in the number of models. By way of example, with Omega, the number of different models, which had totalled nearly 1,600 at the beginning of the 1980s, was slashed first to 850 in 1985 and again to 284 in 1986.[3] In October 1987, senior management announced that around 100 more models would be phased out by the end of the year. The purpose of this operation, which was carried out under the authority of Ernst Thomke, was to refocus the brand on a limited number of models with a strong identity and to accentuate the connotation of high-end luxury attached to this brand. Accordingly, a decision was taken to eliminate gold-plated models and only keep watches made entirely out of noble materials (gold, silver, platinum) or new materials.[4] Reconfiguring its product offering moved Omega back into the black in 1986 for the first time in many years.

Yet this new marketing strategy was not implemented solely for individual brands. In 1990, Nicolas Hayek reorganized operational management at the Swatch Group in order to introduce brand management Group-wide, characterized by brand differentiation and market segmentation – a strategy requiring closer coordination between those responsible for the Group's different brands. Accordingly, the Executive Group Management Board, the body tasked with overseeing the merger and streamlining the company during the 1980s, was restructured, becoming the platform for coordination between brands and the adoption of a global strategy. Two key changes occurred at this juncture: Ernst Thomke left, accompanied by several of his close business partners; and the Extended Group Management Board was established with some 15 members, which was to become a veritable marketing platform (1990). Thomke's departure, after a falling-out with Hayek, marked a paradigm shift from rationalization-driven management to marketing-oriented management. Most of the men who left senior management in 1989–1991 were former managers active within the holding company, like Müller, in charge of finance and administration

(1989), Mangold, Director of Human Resources (1990), Walther, Deputy Finance Director (1990), and Gautier, Communications Officer (1991). Subsequently, these positions were no longer represented on the Group Management Boards. Clearly, the managers responsible for applying Swatch Group policy no longer sat on the Board, which shifted from the operational aspect to strategy.

Thus, in addition to Anton Bally, who took over from Thomke and ran production (ETA), of the eight people appointed to the Extended Group Management Board in 1990, half were brand managers: Walter von Kaenel (Longines), Hans Kurth (Omega), Peter Petersen (Swatch), and Roland Streule (Rado). There was also one representative each of technology businesses (Willi Salathé) and watch finishing manufacturers (Paul Wyers). Finally, two managers from marketing and distribution joined the Extended Board. The first was Raymond Zeitoun, CEO of the distribution firm SMH France since its founding in 1988. He had many years of experience in the French watchmaking market, as he had initially been recruited by Seiko to run Matra Holding Horlogerie in the early 1980s, before founding his own company in 1984, Inthor SA, which distributed Tissot and Rado watches in France, bringing him closer to the Swatch Group.[5] The second new member was a manager with a marketing and communications background, Franco Bosisio, in charge of the Italian market and the Swatch brand in Italy, whose recruitment was a real coup for Swiss watchmaking. This centralization of marketing power made it possible to adopt a global strategy and oversee its implementation on the various markets.

In the years which followed, subsequent appointments to the Board and the Extended Group Management Board confirmed these bodies' role as marketing platforms. Rather than competing with each other, as had been the case previously, Swatch Group brands could now target different, complementary publics. As far as the implementation of this market segmentation strategy is concerned, the acquisition of Blancpain marked a watershed.

The acquisition of Blancpain in 1992: a decisive step

When the Swatch Group purchased Blancpain in 1992, it also bought a legacy of history and tradition. The brand came from the company Rayville SA – Montres Blancpain, a small family-owned firm based in Villeret, in the Bernese Jura, which SSIH acquired in 1961 as part of its strategy for expanding its production capacity by purchasing other enterprises. This allowed Omega to source parts for small calibre precious watches, one of Blancpain's specialties after World War II.[6] However, SSIH gradually lost interest in the brand, selling it in 1983 for the token sum of CHF18,000 to Jean-Claude Biver, a former manager at the high-end watchmaking company Audemars Piguet, and Jacques Piguet, director of the firm Frédéric Piguet SA, one of the rare independent manufacturers of high-end movements, whose annual production at the time was some 50,000 pieces.[7]

The strategy adopted by Biver, whom the Swiss press called a "nostalgia salesman",[8] went in the opposite direction to the entire watch industry: he categorically rejected quartz and built the image of a brand based on tradition and technical excellence, laying the foundations for the brand to move upmarket. He then relocated the company to Brassus, a village in the Vallée de Joux which is the historical seat of the production of ultra-complicated watches. This move to a region which typifies watchmaking excellence and which had also hosted such firms as Audemars-Piguet and Jaeger-LeCoultre ever since their founding in the 19th century confirmed Blancpain's legitimacy, as did its sourcing of movements from the movement manufacturing firm of its shareholder Jacques Piguet. Moreover, by taking the date of 1735 as the date of the company's founding, Blancpain declared itself the "world's oldest watchmaking brand" and rapidly succeeded, selling its watches, often equipped with complications (moon phases, tourbillon, erotic figures, etc.), as objects of tradition or luxury.

Just six years later, Blancpain's turnover was booming (CHF56 million in 1991, compared with 8.9 million in 1985). However, it remained modest in the light of overall watchmaking turnover at the Swatch Group (CHF2 billion in 1991). By taking over Blancpain and Frédéric Piguet, the Swatch Group was looking to acquire new know-how in the field of marketing and the production of complicated pieces rather than to make rapid market share gains. Indeed, this acquisition provided an ideal opportunity to internalize the proven marketing abilities of Jean-Claude Biver – who was appointed first to the Extended Group Management Board (1992), then to the Executive Group Management Board (1993) – and his team, then apply them to the Group as a whole. What is more, Biver was soon tasked with revitalizing the Omega brand, for which he was placed in charge of international marketing (1993), then appointed to the Board of Directors (1997). What is more, purchasing Blancpain gave the Swatch Group its own unit for the production of complicated, high-end movements, a prerequisite for a transition to luxury.

Towards differentiation in terms of brand positioning

In the middle of the 1990s, the Swatch Group embarked upon a policy of differentiating and repositioning its various brands, with a view to improving their complementarity and boosting the company's ability to compete across all market segments. It made a special effort at differentiating the Group's brands of the "affordable luxury" segment, Omega, Longines, and Rado. The aim was to reinforce each brand's own image, thereby generating different yet complementary products targeting different publics. Omega was selected as a luxury product designed to counter Rolex and Cartier on world markets. It went on to become the Swatch Group's main brand.

As for Longines, it was repositioned in a less expensive segment and graphically redesigned as an object of elegance and classicism, which did not compete head on with the Group's other products situated in the same product range, namely, Rado (high technology and modern design) and Glasshütte Original (technical tradition and classicism). For example, since the 1990s, Longines has refocused on sponsoring activities consistent with its image of classic elegance (riding, gymnastics, skiing, tennis), abandoning less refined areas such as Formula 1 racing (1992). Similarly, it reinforced its historical roots with the publication of two works in 1992: one retraced the firm's history, while the other was a reprint of a report on the American watch factories drafted in 1876 by Jacques David, who was technical director at Longines at the time. Unlike Blancpain, which emphasizes the traditional know-how of its watchmaking craftsmen, Longines is part of the historical tradition of *manufactures*, which played a pioneering role in the modernization and industrialization of watch production, an image which is more consistent with its "affordable luxury" products.[9]

Rado developed in a similar fashion to Longines, marked by disinvestment in the brand in favour of Omega. This company, which was founded in 1917, adopted the Rado brand in 1954. It rapidly made a name for itself as one of the leading Swiss watch brands in the Far East and the Middle East. In 1980, Rado was even the sole Swiss watchmaking company to engage in poster advertising in Shanghai.[10] In 1982, it posted a turnover of some CHF200 million for an annual production volume of around 1 million watches, 96 per cent of which were sold in the Far East. One of Rado's hallmarks has been the use of high-tech materials (hard metals, ceramic, lanthanum) for some of its models, ever since the DiaStar watch (1962). As a result, when the Swatch Group was founded, Rado was a major force among the watchmaking firms. Its CEO, Paul Lüthi, was even the sole representative of a brand to sit on the Group's Board of Directors (1984–1989). 'Partly thanks to the restructuring it underwent in 1987, with the personal input of Thomke, Rado clearly remains one of the Group's main brands. However, the Swatch Group's new strategy has turned it into a global brand specialized in futuristic designs. This differentiation has been accompanied by a repositioning to the mid-range segment in order to leave the high end free for Omega, especially in the Far East. Thus, in the 2010s, the new generations of urbanized Chinese are, unlike their elders, more familiar with the Omega brand than with Rado.[11]

In this process of brand differentiation, the use of entertainment celebrities as brand ambassadors was a key component of communication to the general public.[12] As can be seen in other luxury sectors, Hollywood stars have gone from the status of customers to that of brand emblems. Actors, top models, and athletes are now paid to wear watches. However, as Kapferer and Bastien stress, it is essential to "underline the difference between paying a star to appear in an advertisement and using a personality as a brand testimonial user. In the latter case ... you show the extraordinary personality using the product in everyday life."[13] Thus, Hollywood stars are not models who pose wearing their sponsor's watches; rather, they are extraordinary testimonials for the use of a product, like Marilyn Monroe, who said that a few drops of Chanel No. 5 were all she wore to bed.[14] What is more, genuine luxury ambassadors work hard to promote the cultural and charity activities of their brand. They display an interest that goes beyond their advertising value as such.

It was precisely in this context that the Swatch Group began to turn to brand ambassadors, starting by hiring American supermodel Cindy Crawford and movie star James Bond as Omega ambassadors

(1995). Cindy Crawford became actively involved in designing Omega watches for her personal use – as the advertising campaign was at pains to stress – and the new "Constellation" collection became "Cindy's Choice" in Omega's advertising. Similarly, the character of James Bond appeared as an ideal ambassador from this point of view, because he put his Omega Seamaster through its paces in the course of his adventures around the planet.

For the Swatch Group's other brands, positioned between entry level and the mid-range, this new role of brand ambassador was intertwined with the more traditional role of advertising medium. Nevertheless, using celebrities served to reinforce the brands' new image with the public and to differentiate them. In this respect, using ambassadors helped personalize the relationship with the customer and recreate a close link with the consumer. The brands therefore chose ambassadors who best matched their target public. For example, Longines, which is geared to elegance and classicism, chose among its ambassadors for 2009 the former tennis champions Andre Agassi and Stefanie Graf, the Bollywood actress and former Miss World (1994) Aishwarya Rai Bachchan, and the Chinese men's gymnastics team. As for Tissot, which was positioned as a younger, more sports-oriented brand, in 2010 it selected NBA basketball star Tony Parker, motorcycle road racer Thomas Lüthi, and ice hockey player Steven Stamkos.

6
The Major Move into Luxury (Since 1995)

The marketing strategy adopted in the early 1990s was pursued and developed in the years that followed. The main thrusts of this policy were repositioning to the high end; strengthening of differentiation between brands; and major investment in distribution. This means that whereas company growth in the 1980s was based on rationalization efforts within the Group, it was subsequently increasingly driven by marketing. Production systems were even revamped from this perspective – with a shift from manufacturing movements to manufacturing finished watches. The new marketing strategy implemented from the mid-1990s onwards was based on four main thrusts.

The first thrust: repositioning to exclusive luxury

The Swatch Group's repositioning strategy was part of the overall radical change which took place in the luxury industry during the 1980s, with the strengthening of the distinction between *exclusive luxury* and *affordable luxury* or "New Luxury".[1] Whereas the former concerned extremely expensive high-quality products intended for a small social elite – the traditional meaning of luxury – the latter referred to products sold as luxury objects but financially affordable by a broad clientele. This product distinction made it possible to target different segments and pursue parallel strategies based on exclusivism and the democratization of luxury.

The watchmaking industry fitted neatly into this framework, with, for example, the arrival of Richemont (1988) and Moët

Hennessy Louis Vuitton (LVMH, 1999). The Swatch Group tried very early on to position itself at the high end, with, for example, Omega's 1984 launch of a collection called "Louis Brandt", in a reference to the company's founder, which was mainly composed of ultra-complicated mechanical watches – with some of the movements supplied by the firm Frédéric Piguet SA, which the Swatch Group subsequently bought along with Blancpain (in 1992). What is more, this new collection hit the market at a time when Audemars Piguet, a family-owned company which had remained independent, was gradually withdrawing from the Omega retail network, which it had been able to use since 1969. Accordingly Omega saw a need to offer its own high-end products by its retailers. Audemars Piguet was a high end product, complementary to Omega's goods.[2] However, it is difficult and extremely costly to bring a new luxury sub-brand to market, and this attempt was a failure. The Louis Brandt brand was phased out of Omega's collection in the early 1990s.

The other strategy which the Swatch Group adopted to position itself at the high end was to buy up existing firms. Nicolas Hayek was reportedly interested in various renowned firms (Ebel, IWC, Jaeger-LeCoultre, Lange & Söhne), but he would have needed to either take out a large bank loan or broaden the shareholder base to buy them, an approach which ran counter to his desire to be independent from the financial world.[3] Finally, taking Biver's success with Blancpain as a template, the Swatch Group established itself in the high end during the 1990s by acquiring less famous brands and giving them a makeover. It purchased four brands in 1999–2000: Léon Hatot (1999), Breguet (1999), Jaquet Droz (2000), and Glashütte Uhrentriebe (2000). All four had certain characteristics in common: they were all long-standing firms with prestigious names but were not integrated into industrial groups with a global growth strategy.

The first to be purchased, Léon Hatot, was a jewellery company founded in 1905 by a watchmaker of the same name (1883–1953), which had made a name for itself in the interwar period with its Art Deco products, before switching to the production of electric clocks, a field in which the company went on to specialize after the death of its founder.[4] The Léon Hatot brand, which had virtually sunk into oblivion, regained much prestige when stocks of jewels deposited with a Swiss bank for safekeeping when World War II broke out were auctioned off in 1989. The brand was subsequently purchased by the Swatch Group in 1999.

That same year, the Swatch Group snapped up the Groupe horloger Breguet, which consisted of four firms: Montres Breguet SA, for watch production; Breguet SA, for retailing; Nouvelle Lémania SA, for movement production; and Valdar SA, for parts production. This prestigious name is a reference to the watchmaker of Swiss origin who settled in Paris, Abraham-Louis Breguet (1747–1823), whose firm branched out beyond watchmaking after his death.[5] The brand was subsequently used by various investors, including the Parisian jeweller Chaumet (1970), who transferred it to Switzerland shortly thereafter (1976). When Chaumet went bankrupt in 1987, Breguet was taken over by the Bahraini investment fund Investcorp, which restructured the company before selling it to the Swatch Group in 1999. This let the Swatch Group kill two birds with one stone, as it were: acquire manufacturers of high-end watch movements and parts (Nouvelle Lémania and Valdar) and get its hands on an established exclusive luxury brand.

The Jaquet Droz brand, which the Swatch Group bought in 2000, had a profile similar to Breguet. The brand was based on the name of a famous Swiss watchmaker, Pierre Jacquet-Droz (1721–1791), which had been used by various manufacturers before its acquisition by a financial company based in the Cayman Islands, Cupola Venture Partners Ltd, which then sold it to the Swatch Group.[6] Ever since 2000, the brand has been used for the production of limited series of mechanical watches.

Last but not least, the German firm Glashütte Uhrenbetrieb GmbH was a conglomerate of watchmaking companies whose origins dated back to the middle of the 19th century and which came under the control of the East German authorities after 1945. Privatized in 1994, the group was then purchased by two German entrepreneurs, who used the Glasshütte Original brand for classical top-of-the-range watches, and finally acquired by the Swatch Group in 2000.

The same strategy was used for all four companies. The Swatch Group picked up prestigious brands whose main weaknesses were of a marketing-related nature: poor retail networks and the absence of a distinctive graphic line. The Swatch Group defined designs particular to each brand (Art Deco-inspired jewellery for Léon Hatot; graphic lines bringing to mind watches made by Abraham-Louis Breguet; modern design inspired by the watches and clocks made in the 18th century for Jaquet Droz; and classicism and *grandes complications* for Glasshütte Original), incorporated them into its international retail network and turned them into global brands. What is more, the integration of these four companies into the Swatch Group facilitated rationalization in terms of production, as the manufacturing of high-end movements was itself restructured and centralized within the Group during the 2000s.

Finally, the skills acquired with regard to the design and marketing of exclusive luxury watches led to an agreement with Tiffany & Co. in 2007 for the production of watches bearing the name of the luxury American jeweller. However, this cooperation came to an early end in 2011, as Tiffany did not attach enough importance to watchmaking in the eyes of its Swiss partner. The Swatch Group did not abandon its goal of strengthening its position on the jewellery market, but this was not an easy task. The jewellery collections of the Group's watch brands, such as Breguet and Omega, did not do as well as expected. Moreover, Léon Hatot watches, which were repositioned in the jewellery segment in the beginning of the 2000s, proved a failure. Adversely affected by the financial crisis of 2008–2009, they disappeared little by little from most of the Tourbillon shops, which brought together the Group's main luxury brands under a single roof. Launching a competitive jewellery watch is an extremely difficult undertaking: Piaget was one of the few which managed to make a name for itself on this market, but that was back in the 1960s.[7] So the only solution left for the Swatch Group was to buy one of the top names in the jewellery sector in order to control its watchmaking activities, following its disastrous experience with Tiffany. Even if there is a high potential of growth for the jewel watch segment, the cooperation with Tiffany led to numerous difficulties for the development and the sale of watches. For the Swatch Group, purchasing a jewel company appeared since then as a better strategy, because it made it possible to control alone the way to build this new segment. However, after LVMH acquired Bulgari in 2011, there were not many potential targets left for the Swatch Group. As a result, in 2013 the Swiss multinational bought the American jeweller Harry Winston, which had had a small watchmaking division since 1989 (Harry Winston Timepieces).

The second thrust: better brand positioning

The main reason for entering the exclusive luxury segment was to promote affordable luxury brands, primarily Omega. Of course, certain exclusive luxury brands generated large profit margins, estimated at 23 per cent for Breguet and 20 per cent for Blancpain in 2006. However, they accounted for only a small share of turnover. According to Helvea, they represented a mere 12.6 per cent of the Swatch Group watch sales in 2006 (see Table 6.1). Nevertheless, they had significant growth potential, thanks to their integration in a globally organized group.

Their main benefit was indirect in nature: they generated positive image feedback for other Group brands, facilitating their repositioning. The brands which generated the bulk of earnings were those situated in the affordable luxury segment, primarily Omega, which accounted for over one-third of all watch sales in 2006 (33.9 per cent), but also Longines (9.7 per cent) and Rado (9.3 per cent). Their importance reflects the challenge of the democratization of luxury in watchmaking. The global network of luxury shops Tourbillon Boutique, founded in 2001 and numbering a total of 17 shops in 2010 and 21 in 2013, is emblematic of this relationship between exclusive luxury and affordable luxury. These shops stock a full range of exclusive luxury brands as well as Omega and Swatch, which benefit from the boost in terms of image. Moreover, more display space has been set aside for Omega than for the other brands: this is designed to encourage the consumer to dream but end up buying an Omega, in a more affordable price bracket.

As for the mid-range segment, even though its turnover was lower, it was not abandoned. Tissot was even the Group's no. 3 brand in 2006. Moreover, in 1997, the Swatch Group embarked upon the production of low-priced fashion watches for the American clothes designer Calvin Klein. The Group's aim had changed from securing new outlets for the ETA subsidiary, as with the launch of Pierre Balmain watches in 1987, to establishing itself as a key player in the fast-growing fashion watch segment. This involved a strategic diversification towards a market segment in which the Group was not yet well represented. The Swatch Group (95 per cent) and the American clothes designer Calvin Klein (5 per cent) set up a joint venture in Biel in 1997 under the company name of cK Watch Co.

Table 6.1 Consolidated turnover for the various Swatch Group brands, 2006

Name	Date of acquisition	Sales in 2006 (CHF millions)	As a % of the total	Unit price (in CHF)
Bregnet	1999	305	7.8	>10,000
Blancpain	1992	115	2.9	>10,000
Glashütte Original	2000	60	1.5	>10,000
Jaquet Droz	2000	9	0.2	>10,000
Léon Hatot	1999	5	0.1	>10,000
Omega	1983	1,325	33.9	>>,800
Longines	1983	380	9.7	900–3,000
Rado	1983	365	9.3	750–4,000
Union Glashütte	2000	–	–	–
Tissot	1983	395	10.1	300–900
CK Watch	1997	150	3.8	150–1,500
Balmain	1987			
Certina	1983	138	3.5	350–2,000
Mido	1983			
Hamilton	1983			
Swatch	1983	640	16.4	>70
Flik Flak	1987	–	–	–
Endura	1983	–	–	–

Source: *The Swatch Group*. Geneva: Helvea, 2007.

This new firm, which was headed by Arlette-Elsa Emch, started by bringing to market two lines of watches: a high-end line, "Calvin Klein", and a younger, less expensive line, "cK", which was more successful. The first collection was launched in November 1997 at the Geneva Museum of Modern and Contemporary Art and took up the main aesthetic principles of the New York designer (purity and minimalist lines). In this way, cK Watch let the Swatch Group gain a foothold in the fashion watch market. At the press conference held on the occasion of the brand launch, Nicolas Hayek said that "the watches will be mid-range products, with a great deal of research in terms of design."[8] A few years later, on the strength of this success, cK Watch launched a jewellery line (2004) and several models of men's watches.

In the same year as the launch of cK, 1997, the Swatch Group tried to relaunch the Lanco brand on the Chinese market, an attempt which failed but which highlighted the Swatch Group's interest in the mid-range segment. The fact that the Swatch Group's portfolio has always included mid-range brands is worthy of note, because this is not at all the case with the other Swiss watchmaking groups. The mid-level segment is an extremely competitive arena, in which Japanese producers (Seiko, Citizen, Casio) have refocused. For this type of product in particular, mastering production costs is essential, and this explains to a large extent why the Swatch Group is so competitive in this field. Its determination to remain competitive in all market segments has led the Swatch Group to develop new products and new technologies for its mid-range brands, unlike the other Swiss watchmaking companies. By way of example, Tissot has been marketing a touch screen electronic watch (T-Touch) since the year 2000, and in 2013, Swatch announced the fully automated production of a new type of automatic watch with only a single screw (System 51).

But the Swatch Group is not solely represented in the entry and mid-range segments through its Swiss companies. Swatch is a multinational firm which takes advantage of its globalized production system to maintain a presence on the market as a whole. For example, in 2005, it branched out into the booming sector of licensed watch production for the fashion and sports industry through its subsidiary Eterna, specialized in private labels. This was to meet demand for inexpensive fashion watches using the Group's Asian production centres.[9] Some of the household names that have partnered with it since 2006 are the Spanish fashion retailer Mango, the American shoemaker Timberland, and the Japanese firm ASICS, specialized in sportswear and shoes.

The third thrust: strengthening the distribution network

The vertical integration of distribution has been a priority in the luxury watch industry since the end of the 1990s. The aim is twofold: to maintain tighter control over the distribution and sale of products; and to internalize earnings from this activity. The idea is to improve the quality of distribution rather than to increase the number of sales outlets as much as possible. Accordingly, in 2005–2006, the Omega distribution network was restructured on the German, British, and

Japanese markets, where the number of sales outlets was reduced by 20 to 25 per cent.[10] Distribution is an effective tool for reinforcing the exclusive image particular to luxury goods. Moreover, this verticalization has been concentrated on the Swatch Group's high-end brands, with the creation of flagship boutiques.

The importance attached to the distribution and sale of finished watches is reflected by the creation of several dozen new subsidiaries within the Swatch Group: there were 23 such units in 1998, but their number had virtually tripled to 65 by 2009 (see Table 6.2). A look at their geographical and functional breakdown gives rise to three comments. First of all, Switzerland's key role as the driver of this growth is worthy of note. In 1998, the Swatch Group had only a single company specialized in the distribution and sale of watches: Columna SA, a chain of jewellery stores founded in 1943 and subsequently acquired by Longines. Of course, each watchmaking firm within the Swatch Group distributed its own watches directly on the domestic market, but the point should be made that the Group had not entered the retail trade before that time. The nine new companies created in Switzerland during this period were primarily active in the distribution of luxury watches, Blancpain, Breguet, Jaquet-Droz, and Léon Hatot, for example, each having their own distribution company by 2004. The Swatch Group also bought Time Flashing SA in Lucerne (2000) and Distico SA in Ticino (2004), then set up Swatch Retail SA in 2007. Finally, two companies were established to supervise the global distribution of watches from Switzerland – Swatch Group Distribution (2000) and Swatch Group Far East Distribution (2009).

Table 6.2 Distribution and sales companies of the Swatch Group, 1998–2009

	1998		2009	
	No.	%	No.	%
Switzerland	1	4.3	10	15.4
Europe	10	43.5	25	38.5
Asia	6	26.1	20	30.8
Americas	4	17.4	7	10.8
Other	2	8.7	3	4.6
Total	23	100	65	100

Source: Swatch Group, *Annual Reports*, 1998 and 2009.

Second, emerging markets came in for special attention. Up to the late 1990s, the Swatch Group had focused on traditional markets, particularly Western Europe. After 2000, however, it began to set up distribution companies to service the so-called emerging countries, such as India (2001), Mexico (2001), Russia (2002), Thailand (2002), Poland (2004), Taiwan (2004), the United Arab Emirates (2006), and South Africa (2009). The Swatch Group also took interests in retail firms, in particular in Thailand (Wachirapani Co. Ltd, 2002, a company which it subsequently sold around 2010) and in the United Arab Emirates (Rivoli Investments LLC, 2008). Finally, in Saudi Arabia, the Swatch Group took a one-third stake in the distribution company Alzouman General Trading Co. Ltd, which had been managing over a hundred Swatch shops in the country since 1993 (2010).

Against this backdrop, China was in a class of its own, because it had ten trading companies in 2009 as compared with only two in 1998, the Swatch Group Hong Kong and Lanco Watch, a firm which was also based in Hong Kong but was wound up in 1999. The Group's new Chinese companies fell into various categories, encompassing distribution companies (Swatch Group China, Shanghai, 2000; SMH Swiss Watch Trading, Shanghai, 2003) and retail sales (SMH Les Boutiques SA, Shanghai, 2004; Shanghai Rui Jing Retail, 2005; Shanghai Ruihengqi Watch Commerce, 2007; Shanghai Rui Wan Retail, 2007; O Grupo Swatch Ltd, Macao, 2007). The high density of the network of Chinese trading companies is the main reason for the Swatch Group's growing success on this market after 2005 (see Chapter 7).

Third, and finally, a reference should be made to the establishment of trading companies specialized in other activities than distribution alone. Retail sales appear as a major thrust of the new marketing strategy. This was particularly the case for the Group's luxury brands, with the Les Boutiques outlets, which were opened on numerous markets (China, France, Germany, the UK, Italy, Switzerland, United States). Since 2000, the Group's main brands have had their own network of flagship stores, which are designed as not only points of sale but also showcases for brands and venues for events, where global stars and ambassadors make appearances. Last, there has also been a strengthening of ties with companies which control vast retail networks, with some of which joint ventures have been established

(China, Thailand, United Arab Emirates). In the United States, the Swatch Group bought up the Las Vegas-based Borsack distribution chain in 1999 and signed an agreement in 2006 with one of the most powerful and long-standing distributors of luxury watches on the American market, Tourneau (37 stores in the US), to open watch shops in top-end shopping malls.

Likewise, the Swatch Group has consolidated its presence in the retail sales area by setting up a company specialized in watch sales in airports, Tech-Airport Holding SA (2004). This firm manages various types of sales outlet, divided up between the Group's multi-brand shops (Hour Passion), Omega shops, and Swatch Stores. In 2013, Tech-Airport was present in 18 European airports (France, Germany, Ireland, Italy, Switzerland, United Kingdom) and Asian airports (China, Malaysia, Singapore).

The fourth thrust: a production system which serves marketing

The production system has been consistent with the Group's new development policy and has evolved in accordance with the repositioning to the high end. Its main features have been vertical integration of finishing parts manufacturers and the restructuring of quartz movement production abroad.

In general, the production system found a balance at the end of the 1990s. After 1988, the share of Swatch Group employees abroad remained stable despite the cutback in production activities (54 per cent in 1998; 51 per cent in 2005; 54 per cent in 2009), while productivity levelled out at an average of 5,903 pieces per employee in 1999–2004 (see Figure 4.2, Chapter 4). As no production figures have been given since 2005, production trends can no longer be identified. However, one thing is clear: the sweeping production system restructurings of the 1980s and 1990s are over. Rationalization is no longer driving the Group's growth, and changes in the balance sheet fully reflect this. As far as assets are concerned, the category "Property, plant, and equipment", for which figures have been available since 1997, gives a few indications of the Group's investment policy. In 2000, this category accounted for 63.3 per cent of fixed assets and 14.8 per cent of assets as a whole. During the ensuing decade, the value of equipment listed on the balance sheet remained stationary (an average of CHF637.3 million in 2000–2009), but its share fell to 30.1 per cent of fixed assets and 9.0 per cent of overall assets. This does not mean that the Swatch Group stopped investing in production facilities. Annual depreciation for equipment came to an average of CHF139.9 million in 2001–2009 (3 per cent of gross consolidated earnings). This relative drop is due to the fact that the Swatch Group's attention was focused on its policy of investing outside production facilities: in the sales network.

Nevertheless, even though the production system appeared to have reached equilibrium, it underwent a radical change, with the vertical integration of manufacturers of finishing parts like cases, dials, hands, and straps. This strategy should be seen in the light of efforts to reposition to the high end. Moreover, internalizing skills in the field of design has become a major challenge, and the other watchmaking groups are making similar efforts. Until the late 1990s, the Swatch Group had largely relied on independent suppliers. Its main foreign subsidiary for parts was the Italian firm Lascor SpA, a case-making factory which SSIH had acquired back in 1974. Lascor went on to become one of the Swatch Group's primary suppliers of high-quality watchcases. According to Swiss foreign trade statistics, in 2000, Italy was its leading supplier of watchcases, with 9 million pieces, representing 44.6 per cent of total supply. Subsequently, however, its share crumbled (3.2 million cases imported from Italy in 2005, or 18.6 per cent) once China came on stream.[11] The Swatch Group also beefed up its branch network for parts by buying up, one after the other, manufacturers of watchcases (Favre & Perret SA, 1999), hands (Universo SA, 2000), dials (Rubattel & Weyermann SA, 2002, and MOM Le Prélet SA, 2006), pointers for watch dials (Indexor SA, 2007), and oils (Moebius & Fils, 2008), all based in Switzerland. Abroad, the Swatch Group picked up the firm Deutsche Zifferblatt Manufaktur GmbH in Pforzheim in 2006. In addition, the Swatch Group holds a minority stake in Terbival SA, a Swiss watchcase manufacturer.

Yet the Group's acquisition policy has not been limited to finishing parts manufacturers. At the end of the 2000s, there was a consolidation of production facilities for high-end pieces, as the Swatch Group wanted to ensure that it had the means to increase its sales volume in this segment. As a result, Nouvelle Lemania was enlarged in 2002 and stopped deliveries to customers outside the Swatch Group in

2003, aiming to supply only the Breguet brand. Incidentally, it was renamed Manufacture de Haute Horlogerie Breguet (2004). As for Blancpain SA, it announced in 2008 that it was taking over Vica, the company created by Vincent Calabrese, a watchmaking wizard who had developed ultra-complicated watches such as *tourbillon* models for Blancpain during the 1980s.[12] Two years later, in 2010, production of high-end movements was restructured, when Frédéric Piguet SA was bought out by his main customer, Blancpain SA, and Valdar SA was taken over by François Golay SA. This two-pronged restructuring move allowed Blancpain to ramp up its production capacity by verticalizing the *manufacture* and to streamline production with regard to high-end components. Finally, in 2010, the Swatch Group purchased two companies which assembled watch movements, Novi SA, in the Jura, and Tanzarella SA, in Ticino, also with a view to reinforcing the Group's vertical integration.

Thus, these various mergers and acquisitions gave the Swatch Group two integrated *manufactures* in the village of Le Brassus, at the heart of the Vallée de Joux. The Group does not publish any information relating to gross sales or staffing in its luxury watchmaking companies. Moreover, it is not possible to ascertain its links with production networks in Asia. However, by establishing itself and strengthening its position in the Vallée de Joux, the Swatch Group gained access to the most renowned cluster of luxury watchmaking in Switzerland. Indeed, the small community of Le Chenit (population of 4,294 in 2000)[13] hosts not only the Swatch Group but also *manufactures* like Audemars Piguet and Jaeger-LeCoultre, as well as production subsidiaries of the most famous Swiss watchmakers (Patek Philippe, Daniel Roth and Gérald Genta Haute Horlogerie SA, Dubois Dépraz SA, Vacheron Constantin) and numerous subcontractors specialized in the production of specific parts. The dynamic nature of this industrial district is due not only to the maintenance of a technical culture, but also to the utilization of presence in this idyllic region as a marketing resource and to the proximity of the border with France, which provides the necessary cheap workforce for these factories.

The repositioning of the Swiss watch industry to luxury also boosted employment within this cluster: the number of industrial jobs in Le Chenit increased from 2,790 in 2001 to 3,448 in 2008.[14] Assuming that these were only workers in the watch industry, this would amount to 7.4 per cent of the total workforce of the Swiss watch industry in 2001 and 6.5 per cent in 2008.[15] Whereas these workshops and companies clearly play a key role in terms of product innovation and the construction of the idyllic image of Swiss watch *manufactures*, their importance must not be overestimated.

Continued efforts to globalize production

The second characteristic of the production system after 1995 was the continued production of watch movements and external parts in Asia up until 2006. Even though data are extremely fragmented and virtually non-existent after 2004, several indications support this assumption. First of all, the Group's production volume of watches and movements rose from 103.5 million pieces in 1999 to 127.1 million in 2004, an increase of 22.8 per cent over a five-year period. Yet during the same time span, the volume of Swiss exports fell steadily (despite the sharp increase in their value), from 36.7 million pieces in 1999 to 30.7 million in 2004. Even if one considers that all of these exports were Swatch Group products (complete watches sold directly or movements supplied to other firms), this leaves nearly 100 million pieces manufactured outside Switzerland in 2004. The Chinese firm Zhuhai SMH Watchmaking Ltd alone reportedly had a 2004 production volume of some 80–90 million quartz movements.[16] According to Chinese foreign trade statistics, exports of watch movements to Switzerland averaged USD37.9 million annually for 2000–2004, making Switzerland the second largest market (28.7 per cent) for the Chinese watch movement-making industry, behind Hong Kong (58.8 per cent) but well in front of Japan (7.5 per cent). The volume of this trade with Switzerland declined sharply in 2005 (USD11.3 million) and thereafter (less than USD2 million per year for 2006–2009), precisely when the Swatch Group announced its decision to scale back its watchmaking production in China.

Second, the Swatch Group phased out its production and assembly centres located outside Asia during this timeframe. According to annual reports, production was halted in the Virgin Islands (2003) and Brazil (2005), while production sites in Thailand and China were beefed up. In 2004, the Chinese factory Zhuhai SMH Watchmaking Ltd employed 1,800 people, or 8.7 per cent of all Swatch Group employees. However, this was not just a simple shift to Asia; rather, it was a full-blown globalization of production facilities. The increasingly important role of the Americas as an outlet in the early 2000s – 7 per cent of total Swatch Group sales in 1999, rising to 18 per cent in 2004 – no doubt reflects the increasing delivery of quartz movements manufactured in China or Thailand.

However, the Swatch Group announced in 2005 that it was shutting down production units in Asia, without providing any details. This clearly referred to the factory Zhuhai SMH Watchmaking Ltd in Shenzhen. On the one hand, Chinese foreign trade statistics show a steep decline in exports of parts and movements to Switzerland from 2006 onwards. On the other hand, the *China Markets Yearbook* continues to list this firm among Chinese watchmaking companies, but its turnover has fallen sharply and it is mentioned as a company owned by overseas Chinese. What is more, the Singapore-based multinational Junken Technology Ltd, which operates in the field of microtechnology and has been in Shenzhen since 1995, announced in its 2010 annual report that it had purchased Zhuhai SMH Watchmaking Ltd for the production of various types of equipment.

The above changes notwithstanding, China remains one of the main suppliers of external parts (cases, straps, dials) to the Swatch Group and the Swiss watch industry as a whole. However, these business relations do not go through subsidiaries; rather, they consist of simple commercial transactions between purchasers and subcontractors. The watch subcontracting industry has experienced enormous growth in China since 2000, thanks in particular to investments by Hong Kong businessmen.[17] In 2005, they controlled some 20 per cent of all Chinese watchmaking companies, particularly the high-tech ones.[18] These companies supply all of the world's watchmakers, including the Swiss ones.

Before taking up these transactions, we must once again consult the foreign trade statistics. The overall volume of trade does not point to any remarkable trends between 1995 and 2010. Direct exports to Switzerland show strong fluctuations and averaged some USD50 million, i.e. less than the value of Thai or Hong Kong exports to Switzerland, even if a large share of exports from Hong Kong must be considered as indirect exports from China.

The main characteristic of Chinese watch exports to Switzerland relates to the composition of this business, with a major shift in 2004. During the years 1995–2003, China supplied essentially assembled and unassembled movements, as well as cases. In 2004, cases and straps recorded strong growth: to a value of USD16.8 million compared with USD9.1 million on average in 1995–2003. Exports of movements were still high in 2004 but have declined dramatically since 2005, while external parts (cases and straps) recorded a huge increase, peaking at USD50.2 million in 2007. China began to specialize in external parts in the mid-2000s, which also explains the growth in exports of external parts from Hong Kong to Switzerland since then.

Second, overall trends for Hong Kong show an upsurge in growth in the second half of the 2000s. This supplier plays a particular role for the Swiss watch industry. Even though its importance as a parts supplier to Switzerland is comparable to that of Thailand, Hong Kong has specialized in the sale of external parts. Cases and straps amounted to an average of 82.1 per cent of Hong Kong's watch trade with Switzerland between 1997 and 2010. Despite the repositioning of the Swiss watch industry to luxury, which meant that greater importance was attached to watch design and appearance, watch manufacturers aimed at minimizing their production costs for cases and straps, two types of parts that are not included in the legal

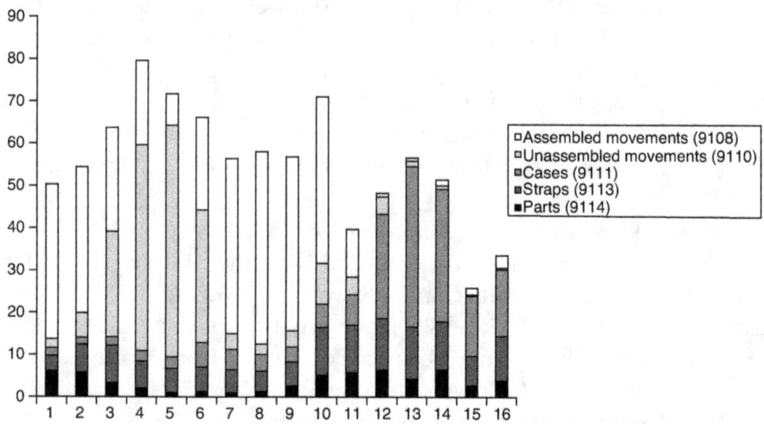

Figure 6.1 Watch exports from China to Switzerland, in USD millions, 1995–2010

Source: Chinese foreign trade statistics, 1995–2010, *Global Trade Atlas*, http://www.gtis.com.

definition of "Swiss Made", the reason for the decline of external parts making in Switzerland. Clearly, the Hong Kong firms that supply Swiss watchmakers with these parts have some production centres in mainland China to take advantage of low wages. Thus, imports of Chinese watch cases by Hong Kong came to USD117.3 million in 2005 and USD256.9 million in 2010.

Switzerland's increasing dependence on China for its supply of parts is particularly visible as far as watchcases are concerned. A reference was made earlier to the move towards verticalization and the acquisition of watchcase makers, a trend which can also been seen in the other watchmaking groups.[19] However, the production of these small firms represents only a very small share of the cases used by the industry, which imports the bulk from China. Whereas the proportion of Swiss watches equipped with foreign cases skyrocketed from 20.8 per cent in 1990 to an average of 60 per cent for 2000–2010, sourcing has to a large extent been refocused on Hong Kong and China.[20] In 2010, these two partners accounted for 82.4 per cent of the volume of watchcases imported by Switzerland, as compared with only 42.8 per cent in 2000 and

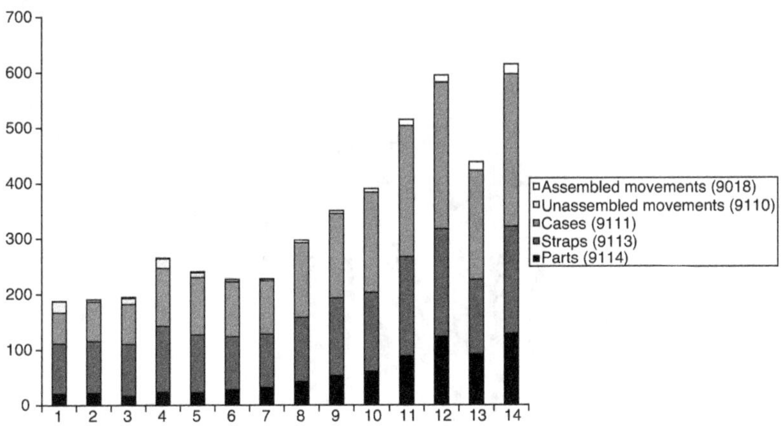

Figure 6.2 Watch exports from Hong Kong to Switzerland, in USD millions, 1997–2010

Source: Hong Kong foreign trade statistics 1997–2010, *Global Trade Atlas*, http://www.gtis.com.

42.3 per cent in 1990. This shift back to China also applies to watch straps, a component which is hardly produced today in Switzerland and which generated annual imports in the neighbourhood of 28.6 million pieces between 1996 and 2012. Over this period, the share of watch straps manufactured by China and Hong Kong rose sharply, from 41.8 per cent to 77 per cent, primarily at France's expense.[21]

China's specialization in activities relating to external parts has been accompanied by a refocusing of activities relating to the production of movements and movement parts in Thailand, where Ronda SA, a manufacturer of quartz watch movements, has also been present since 1989. Ever since the end of the 1990s, Thailand has witnessed a steady increase in its exports of parts and watch movements to Switzerland (USD46.1 million in 1998 and USD135.9 million in 2010). Thus, there is strong complementarity between China/Hong Kong, which have specialized in external parts, and Thailand, which has specialized in watch movements.

Thailand's foreign trade statistics also reveal that a major change occurred in the mid-2000s (see Figure 6.3). Overall watch exports to

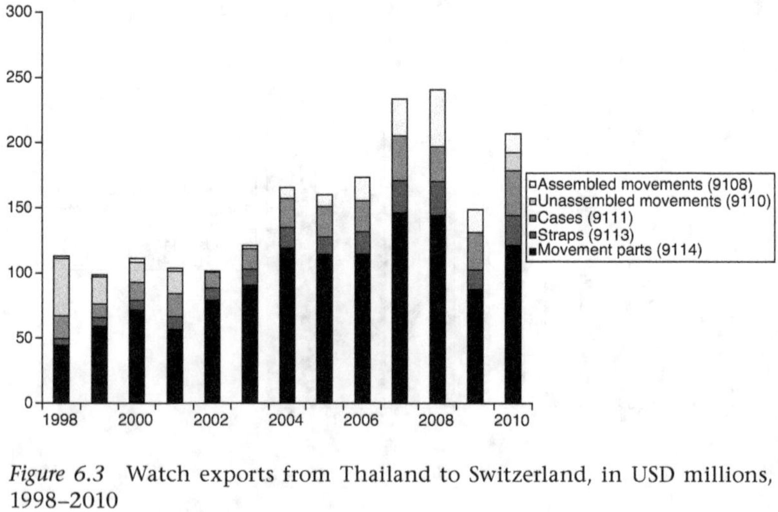

Figure 6.3 Watch exports from Thailand to Switzerland, in USD millions, 1998–2010

Source: Foreign trade statistics for Thailand, 1998–2010, *Global Trade Atlas*, http://www.gtis.com.

Switzerland entered a period of steady growth in 2003: while exports were stable in 1998–2002, they posted considerable growth, peaking at USD240.9 million in 2008. The decline observed in 2009 due to the world financial crisis was obviously only momentary, as new growth could be seen in 2010. Moreover, mention should be made of a structural change in this trade. Even if the volume of technologically simple parts like cases and straps grew steadily, a major change occurred in respect of movement parts. First, parts took on considerable importance, turning into an essential component of this trade (39.2 per cent of the total in 1998 and 56.8 per cent in 2010). Second, unassembled movements (39 per cent in 1998, non-existent in 2003–2009, 6.7 per cent in 2010) gave way to assembled movements (1.6 per cent in 1998, 18.2 per cent in 2008, 7 per cent in 2010), a change which reflects greater technological know-how relating to the design and manufacturing of watch movements in this country.

As for Ronda, the no. 2 Swiss quartz movement maker, it also strengthened its presence in Thailand by opening a new subsidiary in 1990, moving to a new plant in 1999, and finally opening a third production centre in 2009.[22] Yet in 2011, it was the subsidiary of ETA SA, with its 2,600 employees, which remained by far the largest watchmaking company in Thailand (see Table 6.3), well ahead of a Thai-backed firm specialized in external parts (Cosmo Group) and the local subsidiary of the Japanese watchmaking group Casio. In addition, there were four medium-sized firms, almost all of which were foreign-owned.

Asia's central role

Asia occupies a key yet paradoxical position in the Swatch Group's move up to luxury. On the one hand, the supply of movement parts and external parts from Hong Kong, China, and Thailand is increasing despite the fast-paced expansion of a cluster of high-end watchmakers in Switzerland, above all in the Vallée de Joux. This is mainly due to the fact that the Swatch Group has not abandoned the entry and mid-range segments, where manufacturing costs are key competitive factors. On the other hand, Asia accounts for a growing share of the Swatch Group's turnover, rising from 28 per cent in 1999

Table 6.3 Largest watch factories in Thailand, 2011

Name	Foundation	Ownership	Number of employees	Type of activity
Casio (Thailand) Co.	1987	100% Casio Computer (Japan)	800	Watches & electronics
Cosmo Group Public Co.	1967	100% Thai	1,500	Watch dials, luxurious packaging, presentation products
ETA (Thailand) Co.	1986	100% ETA (Switzerland)	2,600	Watch part assembly
Pro-Milling Co.	Unknown	100% Italian	81	Metal parts for watch straps, metal parts for spectacles
Ronda (Thailand) Co.	1989	Ronda 90.57%, Dora Mosset 1.89%, Eris Mosset 1.89%	Unknown	Unknown
Royal Time City Co.	1988	Citizen Watch Co. 80%, Suwan Manufacture 10.22%, Citizen Watch (Hong Kong) Co. 9.78%	498	Watches and watch straps
Siam Dial Co.	Unknown	100% Thai	195	Watch accessories
The Progress Watch Co.	1971	Thai capital 65%, Swiss capital 35%	533	Watch assembly & components (especially cases)

Source: *Factory Directory in Thailand 2012/2013.*

to 35 per cent in 2004 and 51 per cent in 2010 (see Figure 6.4). Yet this steady increase has not primarily been due to sales of cheap electronic movements manufactured in the Group's Chinese and Thai

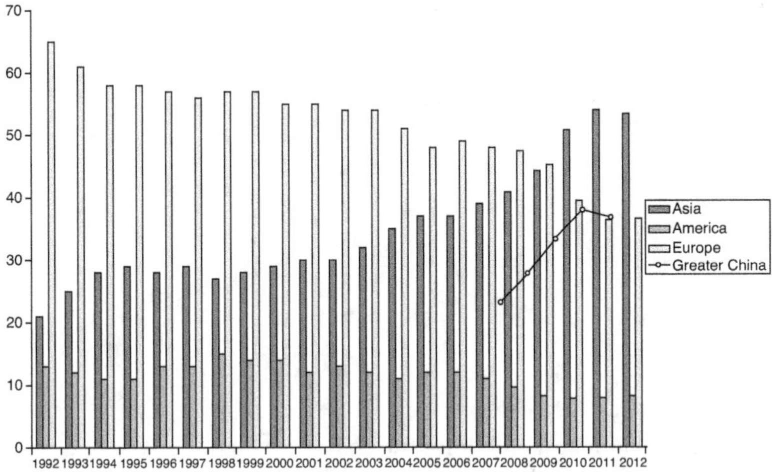

Figure 6.4 Shares of main Swatch Group markets, as a percentage of turnover, 1992–2012

Note: The Greater China region covers China, Hong Kong, and Taiwan. It is also included in the category "Asia".

Source: Swatch Group, *Annual Report*, 1992–2012.

factories. It has clearly been driven by the boom in the sale of luxury watches in this part of the world, as we will see below, using China as an example. The slow but inexorable decline of the European and American markets, together with the rise of Asian markets which has been seen since 2000, clearly reflects the break caused by the move up to luxury. The case study of Omega and the Chinese market, which are taken up in the next two chapters, will enable us to shed light on the main issues at stake in this relationship with Asia.

7
Omega's Choice

Omega no doubt best embodies the changes which the Swatch Group has undergone since the middle of the 1990s. The success of this brand, which accounted for around one-third of Group watch sales in the second half of the 2000s, is emblematic of the Group's new marketing strategy. Omega was picked as the main "affordable luxury" brand, the most profitable market segment in terms of turnover and the one where the major watchmaking groups are fighting it out. Omega's rivals are no longer Japanese watchmaking firms like Seiko, but rather other "affordable luxury" Swiss brands, such as Rolex, Cartier, and TAG Heuer. So the following question comes to mind: how exactly did this brand become a luxury brand in the 1990s?

The arrival of new managers

Omega is one of the oldest and most prestigious Swiss watchmakers. Founded in La Chaux-de-Fonds in 1848 under the name of Louis Brandt & Frère, the firm relocated in 1879 to Biel, where it experienced rapid growth, thanks in particular to the early adoption of mechanized production processes, based on the American model.[1] The company, which employed 600 workers in 1890 and nearly a thousand by 1914, established itself as the main Swiss watch manufacturer. In 1930, the company merged with Tissot into the Société Suisse pour l'industrie horlogère SA (Swiss Watch Industry Corporation Ltd), or SSIH, a holding company whose ownership and management remained in the hands of the Brandt and Tissot families. It remained one of the country's leading *manufactures*, posting strong growth driven by the acquisition of several watch-making companies and the expansion of its brand portfolio, as a result of which its turnover jumped from CHF246 million in 1965 to a highpoint of CHF733 million in 1974, while the volume of watches sold soared during the same period, rising from 3.4 million pieces to 12.5 million.[2]

Despite figures which reflect a certain amount of success, however, the strategic choices made during this period in terms of brand management proved disastrous over time for Omega. In 1970, acting on the recommendation of the McKinsey consulting firm, the SSIH Board of Directors decided to introduce a one-man general management system and brought in Pierre Waltz, who had made a name for himself in the tobacco industry from the early 1960s onwards, in particular as director of Philip Morris Europe from 1967 onwards.[3] The growth strategy which the new management implemented was not based on enhancing the value of the best-known brands, like Omega. Instead, it focused on the continued expansion of production facilities and volumes, inter alia through major investment in the production of cheap watches. Subsequently, SSIH was hard hit by the crisis of 1975–1982, and when the Swatch Group was created in 1983, everything had to be rebuilt in order to reposition Omega as one of the great Swiss watchmaking brands. During the 1980s, the Omega collections were restructured under Thomke's guidance.

Efforts to reposition Omega as an affordable luxury brand got under way in the 1990s, when the Swatch Group decided to make the company its flagship brand on world markets. This policy was implemented by a new team of managers brought in by the director of Blancpain, Jean-Claude Biver, who was appointed Director of International Marketing for Omega in 1993. At the time, the company was headed by a financier, Hans Kurth, and Biver's new team was entrusted with the task of repositioning Omega. Biver recruited several young managers to help him revitalize the brand, most notably Michele Sofisti and Aldo Magada. Born in Italy in 1957, Sofisti had studied geology at the University of Parma and worked as an engineer in this field until 1988, when he joined Ferrari's marketing department. He went on to become a director of Ferrari's German subsidiary before being appointed Director for the Italian market in 1994. The following year, he was recruited by Omega and joined Biver's team.[4] As for Magada, he had trained at the Lausanne Faculty of Business and Economics (HEC Lausanne), like Biver, and at University of California, San Diego, entering Omega as a product manager in 1993 after a career in the pharmaceutical industry and electronics and a four-year stint with Piaget.[5]

In 1997, Omega's internal organization was revamped as the logical outcome of the marketing strategy implemented in previous years. Whereas Biver, who had been promoted to Omega's Board of Directors, relinquished his operational marketing responsibilities to focus on Blancpain, Sofisti was appointed General Manager on Kurth's retirement. Omega was now headed by managers with marketing backgrounds, and when Sofisti left Omega for rival group LVMH in 1999, his job was given to a former Omega marketing director, Stephen Urquhart, who had also worked alongside Biver at Audemars-Piguet and had been director at Blancpain SA from 1997 onwards.

Thus, during the 1990s, Omega's marketing department was a training ground for managers, whose influence was not limited to the Swatch Group.[6] The managers who shared this background were instrumental in moving the entire Swiss watch industry up into the luxury segment. The career of Michele Sofisti is a case in point. After leaving Omega in 1999, he briefly headed up Christian Dior Watches, which was part of the LVMH Group (1999), then came back to the Swatch Group to lead the Swatch brand (2000), which he subsequently left again in Biver's wake (2005). After that, he directed the watchmaking department of the Gucci Group, in which the French luxury group Pinault-Printemps-Redoute, or PPR (now Kering), took a majority interest in 2003, before being appointed General Manager of the Sowind Group (which includes the brands Girard-Perregaux and Jeanrichard) after it was taken over by PPR in 2011. As for Aldo Magada, who was appointed Marketing Director in 1996, he left Omega in 1997 to take over marketing at Ebel, then continued his career in various watchmaking companies (Gucci, Reuge, Breitling).

Likewise, the members of the management team working under Sofisti and Magada shared similar characteristics. Several went on to pursue careers outside Omega and the Swatch Group. The experience they had acquired in terms of communication, advertising, and brand management was put to work for the benefit of other watch brands. For example, this was the case with Béatrice de

Guervain, who worked on this team from 1994 onwards and was promoted to Head of Marketing for Omega on the US market in 1997. When she left Omega, she joined Chopard (1999), Harry Winston (2004), and then Hublot (2010), where she teamed up with Jean-Claude Biver once again. Venanzio Ciampa is yet another example of this new breed of managers. This Italian marketing consultant was appointed Marketing Director for Omega when Sofisti became Managing Director for the brand in 1997. Two years later, Ciampa left the Swatch Group for LVMH before returning to head up the Swatch segment and then founded his own communications agency specialized in the luxury industry and watchmaking, The Promotion Factory, in 2004. The 1997 shake-up of the marketing team also led to the appointment of Yolande Perroulaz as public relations officer and Valérie Bastardoz as Advertising Manager. The former went on to pursue a career within two Swatch Group companies (Longines, Eterna) and the watchmaking divisions of other luxury multinationals (Montblanc, Fendi), whereas the latter joined the banking industry in 2002 after having worked with several Swatch Group brands. These career paths clearly show how significant an impact the marketing managers who joined Omega after Biver's arrival had on the entire Swiss watch industry.

This team introduced a new marketing strategy designed to strengthen the Omega brand's position as an "affordable luxury" product. Communications, which were at the heart of this strategy, were completely revamped. Omega dropped image-driven advertising for message-driven advertising, using new slogans like *Omega – the link between the past and the future* (1994), *Omega – trust your judgement* (1995), and *Omega, my choice* (1997). The brand became a vehicle for images and emotions which tell a story, a process which creates added value and makes it possible to turn a watch into a luxury good. Media events were the cornerstone of this policy. For example, after staying away for seven years, Omega staged a comeback at the Basel Watch Fair in 1993. It used the event as a springboard for its new communication theme, organizing a "space forum", with several famous astronauts in attendance. Omega's participation in the American space programme was used as a key communication tool. The following year, Omega once again caused a stir in Basel by inviting Neil Armstrong and launching a new collection of the Speedmaster model – which had accompanied the American astronauts to the moon in 1969 – to mark the 25th anniversary of the first moon landing. Similar media events were staged when other collections were updated: for example, ambassadors were brought in, like the American supermodel Cindy Crawford or the tennis player Martina Hingis – with whom Nicolas Hayek played a high-profile tennis match in June 1997 – to celebrate the roll-out of the revamped Constellation collection in 1995. The two basic principles of the marketing strategy implemented by this new team of managers were rebranding and the restructuring of distribution networks.

Strengthening brand legitimacy

First and foremost, rebranding was based on strengthening Omega's threefold "legitimacy" – technique, history, and glamour – on which the brand was now based. Omega began by reinforcing its image for technical excellence, which had been tarnished in previous decades by strategic choices designed to boost production and sales volumes.

The chronometer rating certificates awarded by the Contrôle Officiel Suisse des Chronomètres, or COSC (Official Swiss Chronometer Testing Institute), which represent official recognition of the technical quality of watches, show a trend which began in 1960 and which clearly reflects the strategic changes implemented by Omega's management over a period of nearly five decades. The relationship with Rolex, which has remained committed to a management model based on the mass production of high-quality chronometer watches, sheds light on the issues at stake in Omega's repositioning (see Figure 7.1). Up to 1974, Omega registered a growing share of chronometers, which accounted for 10.5 per cent of the company's overall production during the period 1961–1970. Together with Rolex, it was the main chronometer producer: between 1961 and 1974, their shares of the total were 48.9 per cent for Omega and 38.2 per cent for Rolex. After 1974, however, Omega entered a time of crisis which did not seem to affect Rolex. The number of chronometer rating certifications for Omega was in free fall, dropping from 190,396 pieces in 1974 to fewer than 30,000 pieces per year during the period 1985–1989. After the creation of the Swatch Group, restructuring even led to a sharp decline in chronometer rating certificates, which averaged a mere 5,657 pieces per year in 1985–1989. With only

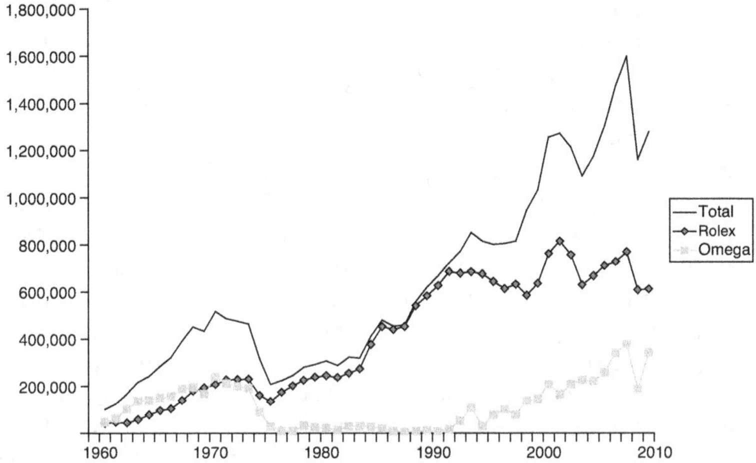

Figure 7.1 Number of official chronometer rating certificates issued, 1961–2010

Source: Annual reports of the Office for the Office for the Inspection of Watches (1961–1973) and COSC (1974–2010).

5.7 per cent of certificates for 1980–1990, Omega abandoned this segment to its rival Rolex, which turned it into a virtual monopoly (89.2 per cent). A new strategy was adopted in the early 1990s to make up the ground it had lost in the chronometry sector. Since 1994, Omega has shown a general upward trend, despite the recession of 2009 (190,516 certificates in 1994; 146,144 in 2000; 342,798 in 2010), enabling it to regain market share (12.7 per cent in 1994, 14.2 per cent in 2000, 26.9 per cent in 2010). Moreover, Omega is not the only brand that has leveraged chronometer rating certification as an instrument for moving upmarket. Other brands have followed in its footsteps and have started to submit pieces to COSC. The share of watches other than Rolex and Omega came to a meagre 6.9 per cent in 1994, but rose considerably to 24.3 per cent in 2000 and 25.3 per cent in 2010. In 2009, the other main brands were Breitling (9.3 per cent), TAG Heuer (6 per cent), Panerai (3.5 per cent), and Chopard (3.2 per cent).

Omega's repositioning as the main Swiss manufacturer of chronometers alongside Rolex was aimed at restoring the brand's technical legitimacy. This strategy was reinforced by the relaunching of a *tourbillon* model (1994) and above all by the use of a new kind of escapement called "coaxial" (1999), developed by the British watchmaker George Daniels, considered one of the world's leading watch craftsmen, followed by the launch of a new Omega movement coupled with this escapement (2007). This marked the first time since the production of watch movements for all Swatch Group brands was centralized within ETA in the second half of the 1980s that movements specific to a brand were developed. Yet this did not imply a return to production within the Omega subsidiary, as it was ETA which developed and manufactured this movement for Omega. To reposition this brand at the high end and strengthen its image for technical excellence, it was necessary to make an exception, on marketing grounds, to the principle of rationalization of production. As far as the consumer was concerned, what mattered was not so much understanding the subtle functioning of this new escapement, which reduces friction and thereby improves movement durability, but above all remembering that Omega was a brand which remained active as far as technical innovation was concerned. Even though it was hard for consumers to grasp the functioning and usefulness of this innovation, it enabled Omega to refocus on the brand's technical excellence. What is more, Omega flagship stores throughout the world displayed the functioning of this new movement on digital screens, adding a technical dimension to the hushed atmosphere of luxury shops. This marketing innovation of a technical nature is an excellent illustration of what Kapferer and Bastien had to say about the role of technology in the luxury industry. They argued that technology "serves only to maintain the objective distance, but on condition that it keeps the dream: there is a potential for imagery and sublimation of the world running through the technology".[7]

Finally, in 2008, the Swatch Group once again used the Omega brand for chronometering the Olympic Games, through its subsidiary Swiss Timing. With one break, when its rival Seiko serviced the Barcelona Games in 1992, the Swatch Group resumed its role of Official Timekeeper of the Olympic Games, but initially used the Swatch brand for its communications tied to this event (Atlanta 1996, Sydney 2000, Athens 2004). Swatch Group choose Omega for the 2008 Beijing Games, because it provided an opportunity to reinforce the brand's image for technical excellence and display a certain historical continuity, as Omega had been responsible for official chronometering of the Olympic Games on several occasions since the Los Angeles Games in 1932. Here as well, the flagship stores were used to get the message across: several of them posted on their walls a list of Olympic Games for which Omega had provided official chronometering, thereby taking credit for a joint effort, because the firm Swiss Timing had included rival firms Longines and Heuer-Léonidas up until the beginning of the 1980s.

Second, Omega strengthened its historical legitimacy, as its communications policy was designed to highlight the brand's long-standing tradition of innovation and technical excellence. Here as well, the strategy implemented is a textbook example of luxury marketing as defined by Kapferer and Bastien, who write that "history gives depth to a brand, and timelessness to its objects".[8] The events organized in 1998 to mark the 150th anniversary of the firm's founding provided an occasion to celebrate its historical roots, with the publication of a book weighing in at nearly 500 pages – *Omega Saga* – stressing in particular the development of sports chronometering, the awards obtained in chronometry contests, and participation in the voyages to the moon, as events making it possible to combine tradition with excellence. A second book, entitled *A Journey through Time*, was published in 2007. Yet the event which marked the high point in Omega's "historical" publicity was without doubt the *Omegamania* auction in Geneva in 2007.[9] The organizer, Antiquorum, the world's premier auctioneer of modern and vintage timepieces, had introduced thematic auctions for prestigious brands and watchmakers starting in 1989 (Patek Philippe, 1989 and 1999; Breguet, 1991; Rolex, 1992; Vacheron Constantin, 1994 and 2005; Cartier, 1996; etc.), which had had a major impact in boosting brand awareness. Thus, the thematic auction devoted to Omega helped reinforce recognition of the brand's historical value and excellence. Omega's historical legitimacy has also been buttressed by the opening of special sales outlets, such as the Omega Vintage shop in Burlington Arcade, London (2008), which sells only vintage models. Another point worth mentioning is that since 2010, internet surfers have been able to order extracts from Omega's archives recounting the historical conditions in which the models in their possession were manufactured.[10]

Finally, Omega's positioning is based on glamour. Up to the middle of the 1990s, the brand's advertising and communications primarily displayed products, stressing the watches' technical qualities and aesthetic appeal. Then the stars started to arrive; supermodel Cindy Crawford was drafted as the first brand ambassador in 1995. However, this radical change was not limited to Omega; it could be observed throughout the luxury industry. Since then, Omega has engaged several Hollywood stars, top athletes, and adventurers as brand ambassadors, but James Bond, who became an ambassador in 1995, no doubt remains the best symbol of this globalization of luxury.

This threefold legitimacy underpinning the brand's repositioning in the mid-1990s allowed Omega to climb back into the ring with its former rival Rolex, a symbol of high-end watchmaking and the world's best-known watch brand. In a similar price range, Rolex has for decades based its communications on the three legitimacies (technique, history, glamour) which Omega is claiming.

Modernizing distribution networks

The second thrust of this new marketing strategy was the reform of distribution systems, with the objective of improving the brand's visibility, not increasing the density of the sales network. When an "ordinary" brand moves upmarket to become an affordable luxury product, this shift must be accompanied by the creation of a distribution system that is consistent with the new image. As Helvea notes in a report on the Swatch Group published in 2007, at the beginning of the 2000s Omega had a vast sales network, but it was completely restructured in the middle of the decade, as a result of which the number of sales outlets was slashed from 4,800 to 3,000, with a 20–30 per cent reduction for Germany, Japan, and the UK.[11]

Along with this drastic downsizing in sales outlets – which mainly affected retailers who had failed to give Omega the visibility it expected – there has been a sharp increase in stores promoting the brand's luxury image. Omega has been one of the main Swatch Group brands that has benefited from the flagship store system. The first such store was opened in Zurich in 2000, in cooperation with an independent retailer, with whom a joint venture was set up. According to the brand's internet site, there were 178 flagship stores throughout the world in 2010, of which 71 were in continental China (10 in Beijing and 10 in Shanghai), 10 in Hong Kong, and 3 in Macao. They have come to play a major role in product distribution. As Nicolas. Hayek stated, "our flagship stores installed in prime locations are advertising in themselves."[12] What is more, they have become a venue for high-society events attended by ambassadors and have become a part of the brand's *storytelling,* along with efforts to educate consumers, who are introduced to the brand's technical excellence by means of animated videos explaining how the coaxial movement works.

8
China: A New El Dorado

Swiss watch exports to China exploded from CHF36 million in 2001 to over a billion Swiss francs in 2010, making this market the fourth largest outlet for the Swiss watch industry, with 6.8 per cent, behind Hong Kong (19.7 per cent), the United States (10.4 per cent), and France (7.2 per cent). If Hong Kong and Taiwan are included, Greater China accounted for 28 per cent in 2010.[1]

Yet watchmaking is not alone in this respect: with few exceptions, today's world industry depends on this new El Dorado and its 1.3 billion inhabitants. In the case of the watch trade, however, the shift to China has gone hand in hand with a remarkable move upmarket: watches in the "affordable luxury" category have been the main beneficiaries of this rapid expansion, and China has played a key role in the process of repositioning the Swiss watch industry to the luxury segment.[2]

Structuring a market

Yet, fundamentally, China is not a new watch market. Swiss and German watch merchants have long been present in China, and watch exports to the country rose sharply after the Chinese Revolution of 1911. In the 1930s, the volume of Swiss watch exports averaged nearly 540,000 pieces per year, corresponding to a market share of 4.7 per cent.[3] However, business relations were hurt by the war, then further undermined when the Communist regime took power in 1949. Initially, the industrial policy adopted by Mao Zedong, even though it focused on big industry, favoured the development of a few watchmaking factories, above all in Shanghai and Canton (Guangzhou) at the end of the 1950s. According to the estimates of the Swiss Chamber of Watchmaking, in 1962, the volume of Chinese watch production ranged between 200,000 and 500,000 pieces. Subsequently, it is estimated to have risen to 1.5 million watches in 1968, then 5–6 million in 1975.[4] The shift to electronic technologies led to a quantum leap in terms of domestic production, which stood at 36.4 million watches in 1984.[5] These watches were primarily meant for the inhabitants of the major coastal cities, then, from the late 1970s onwards, for the entire country, as those living in rural areas began to reap the benefits of the economic reforms adopted after Deng Xiaoping came to power in 1978. Official statistics for the distribution of consumer goods in rural areas point to a sharp increase in purchases of wrist watches during the first half of the 1980s (see Figure 8.1). In 1980, there were only 38 wrist watches for 100 rural households, which means that in nearly two-thirds of households in rural areas no one had a wrist watch. This figure rose

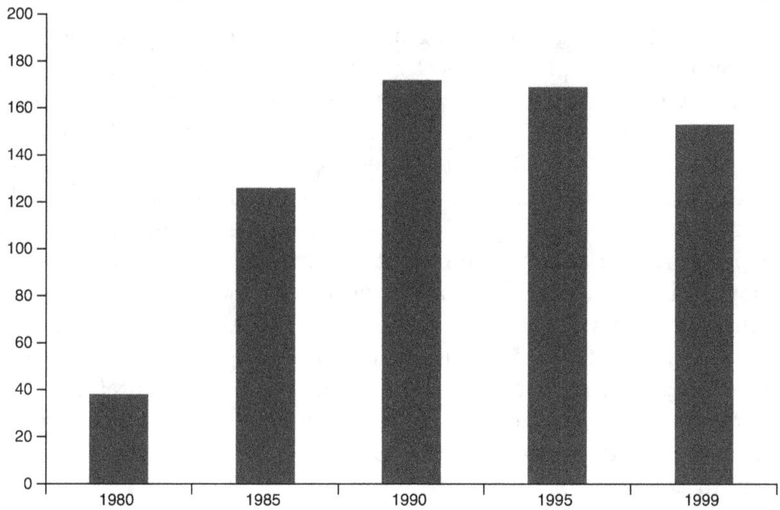

Figure 8.1 Number of wrist watches per 100 households in rural areas, 1980–1999

Note: No figures are available for this index after 1999; nor are any given for urban areas.

Source: *China Statistical Yearbook*, 1980–1999.

to 126 in 1985 and to 192 in 1990, reflecting the rapid spread of watches throughout the country. In the 1990s, the figure fell slightly but remained above 100: watches had become part of China's "material civilization".

As a result, when China opened up to foreign entrepreneurs in the first half of the 1990s, it already had a structured, well organized watch market across the country. This is probably why Swiss watchmakers targeting the entry and mid-range segments did not strike it rich during that decade. The Chinese population had access to cheap domestic watches, and foreign watches could not compete on this market. The Lanco brand, launched by the Swatch Group in China in 1997 for inexpensive utilitarian products manufactured in the country, was a flop. A few years later, the Group experienced increasing success in China with luxury products like Omega and Longines watches, not utilitarian watches.

The sharp rise in Swiss watch exports to China

As we can see with the example of Omega, since the mid-2000s, China has been a key driver of growth in the company. Even though the Swatch Group provides few figures relating to sales by region, we do know that Asia's share of total turnover soared from an average of 28.3 per cent in the years 1995–2000 to 37 per cent in 2005 and 51 per cent in 2010. Of course, these figures include other sectors than just finished watches – in particular, the sale of movements and electronic components produced in Asia. However, this increasing dependency on Asia has come at the same time as the strategy of moving up to luxury and producing finished watches. Moreover, China accounts for a growing share of Swatch Group turnover. Since 2008, the Group has published turnover figures for the market known as "Greater China", which consists of continental China, Hong Kong, and Taiwan: sales for this region represented 23 per cent of consolidated turnover in 2008, 33 per cent in 2010, and 36.8 per cent in 2012.

Thus, the Swatch Group is doing more than simply following the general trend of the Swiss watch industry, for which China emerged as the new El Dorado after 2000. China's accession to the World Trade Organization in 2001 was followed by extraordinary growth in exports to the country, which skyrocketed from CHF36 million in 2001 to 1.6 billion in 2012. Moreover, in terms of market share, China posted steady growth, rising from less than 1 per cent of Swiss watch exports in 2001 and 2002 to 7.7 per cent in 2010. If Hong Kong and Taiwan are included, the Chinese market accounted for 22.4 per cent in 2008 and more than 30 per cent in 2011 and 2012 –shares which are smaller than the corresponding ones for the Swatch Group (see Figure 8.2). Moreover, the Swatch Group managed to strengthen its presence on the Chinese market in 2009, a crisis year for the Swiss watch industry. The Group's turnover for the Greater China region rose from CHF1.3 million in 2008 to 1.4 billion in 2009, then increased further to CHF2 billion in 2010. If this sum comes exclusively from the sale of watches from Switzerland, this would mean that the Swatch Group controlled 34 per cent of Swiss watch exports in 2008 and 46 per cent in 2009. These proportions are no doubt exaggerated by the fact that the Swatch Group's turnover figures for China probably include some non-watch products and watch products manufactured outside Switzerland. Nevertheless, it is no exaggeration to speak of the Swatch Group's domination of the watch

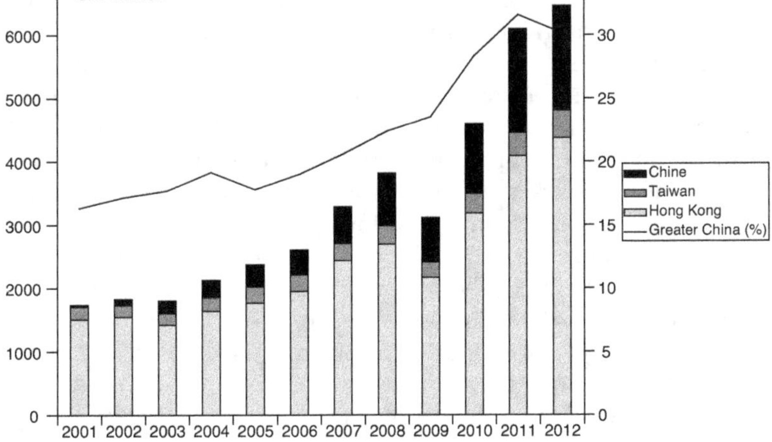

Figure 8.2 Swiss watch exports to Greater China, in CHF millions by region and as a percentage of turnover, 2001–2012

Source: Federation of the Swiss Watch Industry.

trade in China, a position of strength it has maintained with the help of lashings of advertising: the Swatch Group was clearly over-represented in terms of Swiss watchmakers' advertising in the print media in 2010, as Nicolas Hanssens points out.[6]

However, recent business trends in the region point to a market which has become extremely competitive between the various watchmaking groups. For example, even though the Swatch Group has seen its turnover continue to increase (CHF2.6 billion in 2011), it is facing stiffer competition from the other luxury multinationals, such as Richemont and LVMH. In fact, the Swatch Group's turnover for Greater China has been stagnating in relative terms, 44 per cent of the total value of Swiss watch exports to the region in 2010, 42 per cent in 2011, and 44 per cent in 2012.

Reliable data for watch sales on the Chinese market are very hard to come by, but some surveys confirm the dominant position of the Swatch Group's various brands in the country ever since it opened up. The turnover estimated for the various Chinese and foreign watch brands published in the *Annual Book of China Watch Market* reflects this reality (see Table 8.1). According to this source, in 2007, the different Swatch Group brands accounted for over 43 per cent of retail watch sales, with the top three slots occupied respectively by Omega, Rado, and Longines. What is more, Omega alone had a market share of nearly one-fourth, well ahead of its main rival, Rolex (7.1 per cent). These data do not include jewellery brand watches like Bulgari and Cartier. They do, however, clearly reflect the dominant position of Swatch Group watches in China. At the time, there was virtually no competition: apart from Rolex, the main watch brands on the Chinese market were high-volume local products, the major Japanese watch brands, and a few low-end Swiss brands which had begun to specialize in the Chinese market immediately after the war. By the end of the 2000s, however, all the major luxury brands had arrived, drawn by the Chinese El Dorado. Even though the Swatch Group still enjoys a strong position on this market, it is up against some very tough competition.

This success is based on a resolutely proactive marketing policy. Ongoing efforts have been made to set up, develop and densify the distribution network in China. Whereas in 2001 the Swatch Group had only one subsidiary in China, which was responsible for all distribution, Shanghai-based SMH International Trading Co., by

Table 8.1 Market shares of the top 15 watch brands in China, 2007

Brands	**Company**	**Market share (value), %**	**Market share (quantity), %**
Omega	Swatch Group	22.8	1.9
Rado	Swatch Group	8.2	1.1
Longines	Swatch Group	7.6	2.3
Rolex	Rolex	7.1	0.3
Tissot	Swatch Group	5	4.1
Tudor	Rolex	4.7	0.6
Titoni	Swiss company (independent)	4.3	1.9
Rossini	Chinese company (China Haidan since 2008)	2.9	9.6
Citizen	Citizen (Japan)	2.8	6
Enicar	Chinese company	2.8	2.2
Fiyta	Fiyta Holdings (China)	2.7	5.5
Tian Wang	Chinese company	2.6	7.3
Ebohr	China Haidan (China)	2.6	8.2
Casio	Casio (Japan)	1.7	6.3
Ernest Borel	Swiss company (independent)	1.5	1

Source: *Annual Book of China Watch Market 2007*, cited by Michel Chevalier & Pierre Lu, *Luxury in China: Market Opportunities and Potential*, Singapore: John Wiley & Sons (Asia), 2010, p. 78.

2009 it had two distribution companies based in Shanghai (SMH Swiss Watch Trading Co. and Swatch Group China), as well as four retail companies, all based in Shanghai.

The alliance with Xinyu Hengdeli

The lynchpin of this strategy has been the cooperation with one of the largest Chinese watch distributors, Xinyu Hengdeli Holdings Ltd, in which Nicolas Hayek's company took a 6 per cent interest in 2005, which it subsequently boosted to 9.1 per cent in 2010, then more than doubled to 20.4 per cent the following year. Moreover, since 2006, one of the eight members of the Board of Directors, Shi Zhongyang, has been a legal advisor to the Swatch Group, which he entered in 2000. This lets the Swatch Group access internal information and play an active part in the management of the company, which is not the case with rival group LVMH, which also has a share in Hengdeli (6.3 per cent in 2012).

The person responsible for the supremacy of the Hengdeli group on the Chinese watch market is a businessman named Zhang, who has been active in the watch distribution sector since the early 1980s.[7] In 1981, he joined the firm Hua Qiao Co., a state-owned enterprise specialized in the distribution of electrical appliances. Zhang went on to head up the company's watchmaking division from 1982 to 1987, which gave him a great many contacts with foreign watchmakers and domestic retail networks. In 1993, he left the employ of Hua Qiao Co. and started his own business in Hong Kong and China. Four years later, Zwang bought up the firm Beijing Hengdeli Timepieces Ltd, a state company founded in 1957 and specialized in watch distribution and sales in the People's Republic of China, which was privatized in 1990. In addition, through a complex system of cross-shareholdings and shell companies, the Zhang family has stakes in the companies Shanghai Xinyu (founded in 1999), Hefei Xinyu Hengdeli (2000), Tianjin Huichang (2000), Harbin Beiheng Jiefu (2002), Qingdao Xinyu Hengdeli (2002), Liaoning Bao Rui Hang (2003), and Shenzhen Yangguang (2003), all of which are active in the field of watch sales at the local level (shopping malls, department stores, retail shops). By gradually taking over these various companies between 1999 and 2003, the Zhang family extended its influence to retail watch sales. In 2004, this entire conglomerate of companies and holdings was restructured and centralized in the form of a new firm, Xinyu Hengdeli Holdings. The group, which is incorporated in the Cayman Islands and listed on the Hong Kong Stock Exchange, is majority-owned by the Zhang family through offshore companies set up in the British Virgin Islands. Thus, in a few years, the Hengdeli group has managed to carve out a position as the leading watch distributor in China and an indispensable partner for Swiss watch companies seeking to organize the expansion of their sales on a Chinese market which has entered a period of strong growth. This is why the Swatch Group became a minority shareholder in 2005.

In the second half of the 2000s, the Hengdeli group posted strong growth, driven by its commitment to retail activities and the geographical expansion of its sales network, which it broadened in

2006 to include second-tier cities, after moving into the major coastal cities.[8] As the group stepped up its retail activities, which jumped from 34.9 per cent of turnover in 2004 to 77.6 per cent in 2010 (see Table 8.2), its turnover soared from RMB1.5 billion (renminbi = yuan) in 2004 to 8.2 billion in 2010. It has the largest chain of high-end watch stores in China, and has expanded its retail network through a strategy of purchasing companies. The number of its retail shops has increased apace, rising from 65 in 2005 to 350 in 2010 and 452 in 2012. In addition, the Hengdeli group works with over 400 distributors throughout China and has since 2005 reaped the benefits of a strategic partnership with three influential distribution groups (Shanghai San Lian Group Ltd; Shanghai Oriental Commercial Building Ltd; Shenzhen Hengjili World Branch Watches Center Ltd). Together, they are said to control nearly 48 per cent of the Chinese domestic market. Moreover, Hengdeli has based its growth on the geographic expansion of its sales network. In 2005, nearly half of its retail sales were limited to Beijing (18.5 per cent), Zhejiang (18.3 per cent), and Shanghai (10.3 per cent). During the following year, it started to expand outside China, with for example the acquisition of Elegant International Holdings Ltd, which has four shops in Hong Kong. By 2010, continental China accounted for only 59.1 per cent of its turnover. According to the Vontobel Group, Hengdeli is by far the largest watch distribution company in continental China, with a market share estimated at 30 per cent in 2010, far in front of two

Table 8.2 Turnover of the Hengdeli Group, in millions of yuan (RMB), 2004–2012

	2004	**2005**	**2006**	**2007**	**2008**	**2009**	**2010**	**2011**	**2012**
Total turnover	1,518	1,396	2,405	4,578	5,516	5,899	8,216	11,375	12,120
Retail sales	530	637	1,363	3,048	3,742	4,435	6,375	8,589	8,956
Distribution	974	745	1,025	1,439	1,631	1,329	1,661	2,551	5,628
Other	14	15	17	90	143	135	180	235	239
Retail as a %	34.9	45.6	56.7	66.6	67.8	75.2	77.6	75.5	73.9
Distribution as a %	64.2	53.4	42.6	31.4	29.6	22.5	20.2	22.4	24.1
Other as a %	0.9	1.1	0.7	2.0	2.6	2.3	2.2	2.1	2.0

Source: Xinyu Hengdeli, *Annual Reports*, 2004–2012.

Hong Kong-based firms, Oriental, with 6 per cent, and Emperor, with less than 3 per cent.[9] Swiss and other European watchmaking groups do not have any strategic stakes in the latter two distributors.

As is clear from the above, the partnership with Hengdeli appears as an essential component of the Swatch Group's strategy for expansion in China since 2005. Hengdeli's commitment to the retail trade has taken the form of the establishment of flagship stores for several Swiss watchmaking companies, including the Swatch Group. They have joined forces in two major joint ventures, one for the exclusive distribution of Omega and Rado watches (2003), and the other for the management of flagship stores – in particular Omega and Swatch (2007). Finally, the Swatch Group referred for the first time in 2011 in its annual report to the existence of a client accounting for over 10 per cent of its turnover (10.8 per cent), a share which increased slightly the following year (11.1 per cent). Even though the client's name is not mentioned, this clearly refers to the Hengdeli group, as suggested the breakdown of the Swatch Group's various markets.

The creation of flagship stores on the Chinese market is another characteristic of the expansion in China (see Table 8.3). The fact that a virgin market for luxury watches is involved is probably the main reason for this strategy, because there was no need to negotiate with existing distributors and independent retailers, on the one hand, and because no one yet dominated this business activity, on the other

Table 8.3 Flagships stores of the Swatch Group, 2010

Brand	Total	China (including Hong Kong and Macao)	China as a %
Breguet	19	1	5.3
Blancpain	21	5	23.8
Glashütte Original (Watch Atelier)	5	3	60.0
Swatch	> 600	66	10–11
Omega	178	84	47.2
Longines	20	11	55.0
Jaquet Droz	6	2	33.3
Tissot	79	n.p.	?
Boutiques Tourbillon	17	2	11.8

Source: Brands' Internet sites (7 July 2010).

hand. In 2010, this new type of business was widespread in China, particularly for the Omega brand, which had 84 of its 178 flagship stores in China (47.2 per cent), and Swatch, which had 66 (over 10 per cent). Longines and Blancpain flagship stores were also well represented in the Middle Kingdom. This pronounced presence in the territory culminated, in May 2010, in the opening of the Swatch Art Peace Hotel, in Shanghai.

Finally, this densification of sales network was accompanied by a communications strategy targeting Chinese consumers. The Group's various companies signed brand ambassador contracts with local celebrities. Longines was the most committed to this strategy, signing a series of contracts with actress and model Chi Ling Lin (2005), the Chinese men's gymnastics team (1999), and the young male tennis player Tsung-Hua Yang (2009). Tissot was tied to the actress and model Barbie Xu (2005) and to the singer and actor Huang Xiaoming (2010). As for Omega, it partnered with world-renowned actress Zhang Ziyi (2009). In this way, China has been integrated into the Swatch Group's overall glamour policy through several of its sports and entertainment stars. However, as Michel Chevalier and Pierre Lu have pointed out, recruiting local celebrities is not enough to strengthen a brand's luxury image. Several Swiss watchmaking firms, such as TAG Heuer and Chopard, have run into problems because their stars were not well known in Chinese society.[10] However, the Swatch Group brands have managed to choose well known Chinese figures who match the image they wish to promote. Moreover, they get their globalized stars involved with promotional activities which showcase Chinese culture, as, for example, when world-renowned Australian actress Nicole Kidman participated in media events alongside various Chinese ethnic minorities during the Beijing Olympic Games (2008).[11]

9
The Swatch Group's Competitors

Despite Omega's success, the undeniable charisma of Nicolas Hayek, and the Swatch Group's special place in the watchmaking industry, the Group does not have a monopoly on the world watchmaking market. There are a host of brands outside the Swatch Group, including the only watch brand that is consistently ranked in the 100 Best Global Brands of the American magazine *Business Week*: Rolex (68th place in 2009).[1] In order to better highlight the Swatch Group's place on world markets and the sweeping reforms of its production system and marketing strategy, it is useful to take stock of the competition and sketch out the primary characteristics of its main rivals.

Yet the virtual absence of data on the size of world markets and on the production of the various watchmaking companies makes this kind of exercise difficult or even random. The few global analyses which do exist are market studies conducted periodically by financial institutes, such as the Vontobel Group or the firm Helvea, for their investment clients. Naturally, these analyses are very approximate, as the various watchmaking groups generally publish only consolidated turnover figures and do not provide information on the volume of their sales or on the different brands. Fortunately, these market studies do exist. They make it possible to identify certain broad tendencies and highlight the major issues in competition worldwide.

Two types of approach may be used to tackle the question of the relationship between the Swatch Group and its competitors. On the one hand, it is possible to concentrate on the watch brands themselves and to draw up a ranking of the largest in terms of turnover. On the other hand, one can also look at the groups themselves, as each generally has a portfolio of brands with various characteristics. These two approaches are presented briefly in this chapter.

The main watch brands

Vontobel's 2011 analysis of the watchmaking world provides a ranking of the world's leading watch brands, according to their approximate turnover (see Table 9.1). Three comments come to mind. First of all, with the exception of Cartier, this ranking covers only brands of Swiss origin. What is more, the famous Parisian jeweller has sourced its watches from Swiss companies since before World War II and has even had its own production centre in La Chaux-de-Fonds since 1972, which means that its watches are Swiss products, even though they are sold under a French luxury brand.

Second, the point should be made that the Swatch Group's rivals have only a single brand, or two in the case of Richemont, in this ranking, whereas the Swatch Group has five of the ten leading watchmaking brands, which underscores both the success of the brand repositioning strategy implemented during the 1990s and the depth of its presence on the market. What is more, it has another brand in the top 20, Rado (in 16th place with CHF440 million).

Table 9.1 Approximate turnover of the top 10 watch brands, in CHF millions, 2011

Brand	Group	Turnover
Rolex	independent	4,500
Cartier	Richemont	2,380
Omega	Swatch Group	2,230
Patek Philippe	independent	1,150
Longines	Swatch Group	1,120
TAG Heuer	LVMH	980
Tissot	Swatch Group	960
Swatch	Swatch Group	720
Breguet	Swatch Group	720
IWC	Richemont	690

Source: *Watch Industry*, Vontobel Equity Research, 2013.

Richemont also has a diversified portfolio with several brands in the CHF350–650 million zone (Jaeger-LeCoultre, Piaget, Vacheron Constantin, and Panerai), whereas LVMH is dependent on TAG Heuer, as the watchmaking turnover for Hublot and Bulgari are estimated at only CHF400 and 300 million, respectively.

Third, and finally, the bulk of these brands are situated in the luxury segment, and more particularly the "affordable luxury" subsegment. Patek Philippe and Breguet are clearly watchmakers in the "exclusive luxury" segment, that is, extremely expensive goods which are primarily aimed at a small elite characterized by its very high net worth. Apart from Patek Philippe and Breguet, they do not occupy the high end of the ranking. Rolex, Cartier, Omega, and TAG Heuer are all "affordable luxury" brands. They are watches which are situated in more or less the same price bracket and which embody similar values (technical excellence, historical tradition). Each belongs to one of the major watchmaking groups competing to gain control of this market segment.

Here, the Swatch Group stands out, not only because it owns five of the top ten watch brands but also because it is present on all market segments, from entry (Swatch) through mid-range (Tissot) and "affordable luxury" (Longines, Omega) right up to "exclusive luxury" (Breguet). As a result, its position as the world's leading watchmaking group is based as much on the depth of its presence as it is on combined clout. It is the only group with a universal vocation, as the other major watchmaking firms generally specialize in a specific market segment. This being so, approaching the world market by analysing the main companies active in this field also highlights the Swatch Group's general character (see Table 9.2), as the other eight groups with market shares of more than 2 per cent are either companies specializing in luxury or companies targeting the mass market (mid-range).

Groups in the luxury industry

The Swatch Group's main competitors are multinational companies operating in the luxury industry. First, there are companies specialized in watchmaking, such as Rolex and Patek Philippe, which moreover occupy the first and fourth slots as far as watch brands are concerned. La Société des Montres Rolex SA also has a

Table 9.2 The world's main watchmaking groups, 2012

Group	Country	Watch sales, CHF millions	Share of world watch market (%)	Watches' share of turnover (%)
Swatch Group	Switzerland	6,955	18.3	89
Richemont	Switzerland	5,960	15.7	49
Rolex	Switzerland	4,500	11.8	100
Fossil	USA	1,970	5.2	75
LVMH	France	1,785	4.7	5
Citizen	Japan	1,490	3.9	51
Seiko	Japan	1,295	3.4	43
Patek Philippe	Switzerland	1,150	3	100
Casio	Japan	800	2.1	25
Other			31.9	

Note: The heading "other" covers groups with less than 2 per cent of the world market.
Source: *Watch Industry*, Vontobel Equity Research, 2013.

sub-brand, Tudor, launched in 1946, for significantly cheaper models. Nevertheless, it remains the archetypal manufacturer of "affordable luxury" goods: high-quality watches which are mass produced (to the tune of an annual volume of some 800,000 pieces in 2000–2010) and sold as objects of social distinction. It qualifies as a *manufacture* insofar as it produces its own movements internally, while probably relying to some extent on the Swatch Group when it comes to sourcing specific spare parts like springs.[2] As for Patek Philippe SA, it is also an independent *manufacture* with its own facilities for designing and manufacturing watches. Unlike Rolex, however, Patek offers "exclusive luxury" products aimed at affluent customers. Even its entry-level model, the Calatrava, sells for several tens of thousands of Swiss francs, making it a product intended for a very different clientele from that of Rolex, Omega, or Cartier. Furthermore, even though its production volume has risen sharply, doubling from some 10,000 pieces in 1980 to 20,000 in 2000, Patek Philippe is not an industrial watch manufacturer like Rolex or the Swatch Group.[3]

In addition to these two manufacturers specialized in high-end watchmaking, there are also the two main multinationals of the luxury industry, la Compagnie financière Richemont SA and LVMH. These are not watchmaking firms as such, but rather holding companies present in the various luxury fields, such as fashion, jewellery, and food. Watches' share of their respective turnover amounted to only 49 per cent and 5 per cent in 2012. These two major players in the global luxury industry were founded in the 1980s. Their expansion has been driven by a policy of buying up companies active in these various sectors – including watchmaking – whose strong growth is underpinned by the financial and marketing infrastructure of these groups organized on a worldwide scale.

More recently, one could also mention the intervention of the world's third largest luxury group in the watchmaking industry, Kering (formerly Pinault-Printemps-Redoute, PPR), with a watchmaking turnover which Vontobel estimated at CHF250 million in 2012, or some 0.7 per cent of the world market. Thus, we are talking about a small player in the watchmaking business, but one whose brands have a high potential for growth. This French group specializing in large-scale distribution entered the luxury industry in 1999 when it took over the Gucci Group. This Italian fashion company had itself branched out into watchmaking at the beginning of the 1970s, when it entered into a relationship with businessman Severin Wunderman. After Gucci purchased Wunderman's watchmaking firm based in Lengnau and renamed it Gucci Timepieces in 1997, Wunderman went on to acquire the prestigious watchmaking firm Corum a few years later. PPR also purchased Bédat & Co. in 2000, which it subsequently sold in 2009 to Luxury Concepts, an enterprise based in Malaysia which also owns the brand "Armand Nicolet Tramelan". Just before it sold Bédat & Co., PPR took a stake in Sowind (23 per cent in 2008), a company which owns the brands Girard-Perregaux and Jeanrichard, then acquired a majority holding in 2011. Taking over the Girard-Perregaux *manufacture*, which produced some of its own movements, gave PPR a centre for the production of watches for the group's other luxury brands: by way of example, since 2007 Girard-Perregaux has made watches for the French jeweller Boucheron, a firm which also belongs to PPR.

Finally, the family-owned French firm Hermès also markets watches. However, these account for only a negligible share of its turnover (5 per cent in 2012) and Hermès' share of the world watch market was only some 0.5 per cent in 2012, with a turnover estimated at CHF207 million. Hermès has sold watches since before World War II but used to rely primarily on prestigious subcontractors, such as Movado. In 1978, the group founded La Montre Hermès SA in Biel, in order to have its own manufacturing centre. Moreover, ever since the second half of the 2000s, the Hermès group has followed a strategy of investing in companies which produce spare parts and movements in order to secure its supplies, by acquiring a holding in the watch movement producer Vaucher Manufacture, in Fleurier, in 2006, followed by a stake in luxury watchcase maker Joseph Erard SA, in Noirmont, in 2011. These twin moves suggest that the brand intends to develop its watchmaking business.[4] However, despite the arrival of several all-round luxury companies in watchmaking and the proactive commitment some have displayed to this branch, only Richemont and LVMH have managed to establish themselves as major players in this sector.

The two giants of the luxury industry: Richemont and LVMH

Richemont is a group which came into being as a result of an upmarket diversification of the assets of the Ruppert tobacco manufacturing family (British American Tobacco, or BAT) into the luxury segment in 1988. Richemont's purchase of Cartier may be viewed as the group's arrival in the watchmaking industry. It soon expanded its presence in this sector by acquiring several other companies, such as Baume & Mercier (1988), Piaget (1988), Vacheron Constantin (1997), Officine Panerai (1997), Roger Dubuis (2007), IWC (2000), Jaeger-LeCoultre (2000), and Lange & Söhne (2000).

Despite the breadth of this portfolio of brands, they are all positioned in the luxury field. Buying Cartier, along with certain *manufactures* like Vacheron Constantin and Jaeger-LeCoultre, has given Richemont the production facilities it needs to make movements and watches, even though it has no doubt continued to rely heavily on ETA SA and the Swatch Group for basic calibres and various components. The group has also developed a policy of acquiring subcontractors operating in the watch finishing field, as reflected, for example, by its purchase of watchcase manufacturer Donzé-Baume SA in 2007.

The annual reports published by Compagnie financière Richemont SA make it difficult to accurately judge the group's performance on the watchmaking market. The use of various currencies (pound,

Swiss franc, euro) and the classification of brands (as jewellery, watchmaking, accessories, etc.) do not give a clear picture of how the group's watch brands have performed. For example, Cartier watches are classified in the jewellery category – which also includes Van Cleef & Arpels jewels – whereas Montblanc watches are to be found in the "writing instruments" category. Despite this methodological difficulty, it is clear that the group has posted very strong growth overall, with total turnover more than doubling from €2.3 billion in 1999 to 5.2 billion in 2010 (see Figure 9.1). The share of watchmaking, listed as a separate category until 2002, averaged 44.7 per cent for the years 1999–2002. Ever since 2003, the respective shares for jewellery and watches have remained very stable. Thus, watchmaking has contributed to the company's overall growth. The financial crisis of 2007–2008 had virtually no impact on the performance of the group, as it did not rely heavily on the US market. Indeed, Asia accounts for roughly half of Richemont's sales.

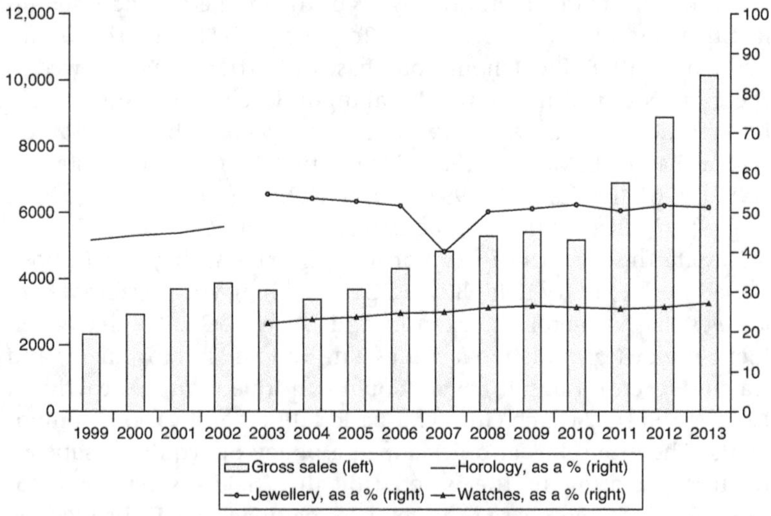

Figure 9.1 Turnover of the Richemont Group, in millions of euros and as a percentage, 1999–2012

Note: Fiscal years run from 1 April to 31 March.

Source: Compagnie financière Richemont SA, *Annual Reports*, 1999–2013.

As for the group LVMH (Moët Hennessy Louis Vuitton), it invested in watchmaking in 1999, with the creation of a "watches and jewellery" division following the acquisition that same year of three renowned Swiss watchmaking firms: Zénith, TAG Heuer, and Ebel (sold on in 2004 to the American group Movado). LVMH subsequently rounded out its portfolio of watch brands by acquiring Hublot in 2008 and Bulgari in 2011. Like its rival Richemont, the LVMH group has only watch brands positioned in the luxury segment, the main such brand being TAG Heuer, which accounted for some 45 per cent of LVMH watch sales in 2010.

Moreover, in 2001 the group inaugurated les Ateliers Horlogers SA, in La Chaux-de-Fonds, a company specialized in supplying watches to the group's other luxury brands, such as Louis Vuitton and Christian Dior. In addition, Zénith and TAG Heuer have their own production centres. Nevertheless, LVMH has continued to depend on ETA and the Swatch Group for its supply of movement blanks and watch parts. Initially, efforts were focused on the acquisition of subcontractors operating in the watch finishing field, with the purchase of watch strap maker Artelink, in Italy (2001), as TAG Heuer had controlled the dial maker Cortech SA since 1992. Finally, it was only recently that the watchmaking companies of the LVMH group set out to design and manufacture their own calibres, TAG Heuer and Bulgari both announcing launches of in-house calibres in 2010.

Unlike Richemont, the watchmaking sector is of minor importance to LVMH. For the years 2000–2010, the "watches and jewellery" division accounted on average for only 4.6 per cent of group turnover, which has recorded strong growth from the mid-2000s onwards. However, investment in the development of own calibres plus the purchase of Bulgari show that LVMH does not intend to abandon the luxury watch industry to its competitors.

The main characteristic of the LVMH group in comparison with the Swatch Group has been its greater dependence on the American market, a legacy of TAG Heuer, a brand well established in the United States but relatively unknown in Asia from the 1960s onwards. In 2004, whereas Nicolas Hayek's company was generating only 11 per cent of its turnover in the United States, the country still accounted for 22 per cent of watch and jewellery sales for LVMH. The latter's share even increased subsequently, to 25 per cent in 2007. This no doubt explains why the financial crisis wrought such havoc in the watchmaking division of the group, whose overall turnover fell only slightly. The previous heavy reliance on the US market gave way to a gradual shift towards China. Although Asia's share was a solid 26 per cent in 2002 – as against 30 per cent for the Swatch Group – this is primarily due to a strong presence in Japan (14 per cent), where LVMH developed a policy of aggressively expanding its distribution network: the number of LVMH watch and jewellery shops in the archipelago jumped from 15 in 2002 to 32 in 2004, as compared with only 7 in the rest of Asia that year. Subsequently, the focus shifted to China, when LVMH took a stake in the Hengdeli distribution company in 2006 (however, the group has no representative on the Board of Directors, unlike its rival, the Swatch Group). Subsequently, Asia's share grew apace, from 28 per cent in 2007 to 33 per cent in 2010 and 40 per cent in 2012, while the US market collapsed (12 per cent in 2012).[5] This repositioning to China, especially for the TAG Heuer brand, combined with the acquisition of Bulgari, has resulted in strong growth since 2010, making LVMH an increasingly serious competitor for the Swatch Group (see Figure 9.2).

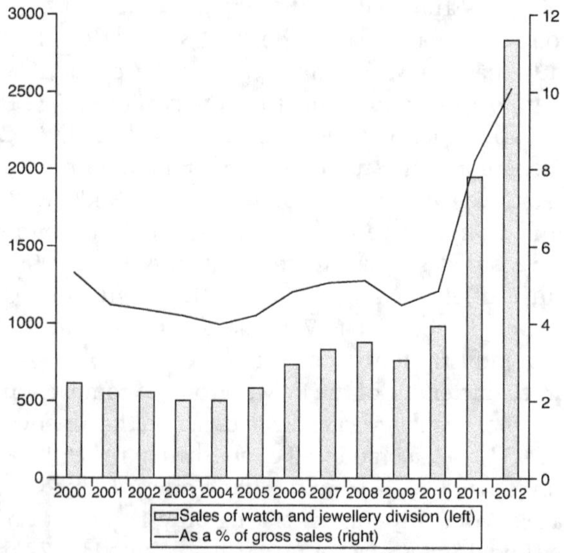

Figure 9.2 Turnover of the LVMH Group's watch and jewellery division, in millions of euros and as a percentage of total group turnover, 2000–2012
Source: LVMH, *Annual Reports*, 2000–2012.

Mid-range watch manufacturers

The classification of the main watchmaking groups by market share includes four companies positioned in the mid-range segment, that is, relatively inexpensive watches (a few hundred francs), generally quartz models, which are mass-produced and distributed. This is a segment in which the Swatch Group is present, in particular with its brand Tissot. The Swatch Group's main challengers in this product range are foreign groups: the Japanese watch manufacturers Seiko, Citizen, and Casio, as well as the American group Fossil.

Having experienced phenomenal success on world markets from 1960 to 1985 thanks to the mass production of high-quality mechanical watches followed by quartz watches, the Japanese watchmaking firms entered a period of crisis in the late 1980s. Their problems were due not only to the improved competitiveness of Swiss watchmakers but also to difficulties particular to this industry. The yen's appreciation following the Plaza Agreements in 1985 had a considerable impact on Japanese industry as a whole, as most manufacturers, including watchmakers, made a massive push to relocate their production centres in South-East Asia owing to the yen's sharp rise. At the same time, these companies adopted a policy of diversifying into electronics (Seiko, Casio) and machine tools (Citizen), that is, areas offering the best opportunities for growth. As a result, by 2010, watches accounted for only a small share of the activities of these three groups, Citizen remaining the most focused on this field of historical activity by means of a strategy favouring the mass production of cheap electronic movements to the detriment of finished watches.

The rise of the American group Fossil

The company Fossil Inc., which is headquartered in Texas, has become the world's leading watchmaking group specialized in fashion watches.[6] Even though the US has a long-standing, rich watchmaking history, featuring such firms as American Watch (known under the name of Waltham Watch), Hamilton Watch, and Bulova Watch, Fossil does not have the slightest connection to either American watchmaking know-how or any other watch manufacturer. The company is emblematic of the new fashion watch producers, and was founded in Dallas in 1984 under the company name of Overseas Productions International by the brothers Tom and Kosta Kartsotis.

Initially, the firm stemmed from a pure marketing project. At the time, Kosta Kartsotis was working as an executive for the Dallas-based department store Sanger Harris, an experience which not only gave him extensive knowledge of the American distribution system but also enabled him to see the profit opportunities in selling watches produced in Hong Kong in the United States. As a result, in 1986, the Kartsotis brothers launched watches bearing the brand Fossil on the American market, whose design and identity were reminiscent of the America of the 1950s. Their marketing gambit soon paid off, and the company's turnover skyrocketed from USD2 million in 1987 to 32.5 million in 1990, the year in which a second brand, Relic, was launched. These initial successes encouraged the two brothers to develop their firm. They opened its share capital up to new investors in 1993 and adopted a strategy featuring three main characteristics: control of production; diversification into fashion brands; and integration of distribution.

Securing the sourcing of watches was one of the first issues the young company had to address. In 1992, it acquired a firm in Hong Kong, which took the name of Fossil (East) Ltd and was responsible for supplying the US headquarters with watches. Fossil also purchased stakes in several other Hong Kong firms, such as Amazing Time (60 per cent), Pulse Time Center Company (60 per cent), and Trylink International (51 per cent). Its ties to these different companies brought it close to one of the world's leading watchmaking centres. According to the financial reports published by the group, in 2000, Fossil sourced watches from a total of 38 watchmaking companies based in Hong Kong and nearby China, a figure which had risen to 48 by 2011.

Next, Fossil adopted a policy of diversification into fashion, in terms of both products and brands. As far as products were concerned, Fossil launched a variety of fashion accessories (leather goods, sunglasses, jewellery) in the 1990s, then a line of apparel in 2000, followed by a shoe collection in 2008. This diversification made it possible to strengthen Fossil's brand identity and its roots in the fashion world, given that it had no historical legitimacy as far as manufacturing fashion goods was concerned. Yet despite this diversification, watches accounted for the overwhelming majority of turnover: their share averaged 68.1 per cent in the 2000s, before entering a moderate growth phase and increasing to 74.9 per by 2012.

Ensuring that watchmaking continues to represent a significant share of turnover is the second thrust of the diversification strategy implemented by Fossil during the 1990s, with the production and sale of licensed goods for various fashion brands. The main contracts have been with the brands Emporio Armani, DKNY, and Diesel, for which the company has produced watches selling in the USD100–200 range. This cooperation was made possible through the experience which Fossil had acquired and its mastery of a production system divided between the United States (marketing) and Hong Kong (manufacturing). This strategy enabled it to become a fully fledged fashion watch group and helped reinforce the image of watches sold under its own brand as fashion goods, as they were showcased in retailers' windows alongside watches bearing the above-mentioned prestigious brands. Moreover, in the years that followed, Fossil broadened its cooperation with other fashion brands and designers. In 2004, it signed agreements with designer Michael Kors and with Marc Jacobs International. The following year, it negotiated a licence to manufacture Adidas watches, which led to the launch of a watch collection featuring a young, sporty design. Subsequently, Fossil joined forces with Karl Lagerfeld to launch watches bearing the world-famous designer's name in 2011. Next, in 2012, the American group bought up the company Skagen Designs Ltd, a small firm founded in 1989 by a couple of Danish designers based in the US, who had launched the Skagen Denmark watch brand in 1992.

Fossil's expansion into the field of fashion watches has been accompanied since the turn of the century by a strategy to enter a much more exclusive sector of contemporary horology: the manufacture and sale of Swiss watches. Thus, in 2001, Fossil founded the group Swiss Technology Holding in Biel and acquired the Zodiac watch brand, whose origins date back to the late 19th century. At the same time, Fossil took over three Biel-based watchmaking firms active in various stages of the watch manufacturing process: Montres Antima (assembly), Synergies Horlogères (watch design), and Méliga Habillement Horloger.[7] The takeover of Zodiac and the few companies around it reflected a twofold aim: first, to bring a Swiss brand with a wealth of history and tradition into its portfolio; and second, to set up production facilities – even though Zodiac clearly depended on outside suppliers for its movements – enabling it to offer its clients watches bearing the "Swiss Made" stamp. Moreover, in 2001, Antima signed a licensing contract with Burberry for the production of Swiss watches. In 2006, Fossil also launched a Swiss Made collection for Giorgio Armani, with watches sold exclusively in the Italian designer's boutiques, whereas the watches for Emporio Armani, the fashion group's second brand, which are geared to the mass consumption market, are made in Hong Kong and distributed directly by Fossil.

Finally, the third driver of growth within the Fossil group since the end of the 1990s has been increasing control over distribution, which is characterized by the integration of distribution into the company, and has been accompanied by the growing internationalization of sales outlets. In 1998, the first year for which detailed data are available, Fossil was a company which still relied heavily on the US market and its department stores. Sales abroad accounted for a modest 29 per cent of turnover, whereas the major American department stores on their own generated 60.8 per cent of total sales.

Between 1999 and 2012, business growth was primarily based on the development of direct sales via the group's website; the establishment of an increasingly dense network of shops; and the purchase of distributors. The year 1995 saw the launch of Fossil flagship stores, and the first such store outside the country opened in 2005. Their number then increased rapidly, from 39 stores in 2000 to 113 in 2008 and 245 in 2012. This network of sales outlets managed directly by the group is a growing source of profit, together with Fossil's webpage. Direct sales' share of turnover soared from a meagre 9 per cent in 1999 to 25 per cent in 2012. During the same period, the group scaled back its reliance on the major US department stores, whose share of group turnover had fallen to 41.2 per cent by 2004.

Moreover, in the mid-2000s Fossil adopted a strategy of direct control over the distribution of its products. One after another, the American firm bought up distribution companies in Taiwan (2005), Sweden (2005), and Mexico (2006). In December 2012, it acquired its third largest distributor, Bentrani Watches, active in Latin America. It also set up its own distribution subsidiaries in Shanghai (2006), India (2007), and South Korea (2007).

As for the internationalization of sales outlets, this has been primarily characterized since 2000 by a decline in the American market in relative terms. Without counting direct sales (internet and the network of flagship stores), America's share of turnover fell from 66.4 per cent in 2000 to 50.6 per cent in 2012. Over the same period, the European market posted a sharp increase (from 21.9 per cent to 32.5 per cent), while the Asia-Pacific region experienced more modest growth (from 11.7 per cent to 16.9 per cent).

The development strategy which Fossil has followed since the middle of the 1990s has made it a success story. The American group's turnover exploded from USD161.1 million in 1994 to 504.3 million in 2000 and 2.9 billion in 2012 (see Figure 9.3). That same year, Fossil's watch turnover hit USD2.1 billion, a level close to the luxury group LVMH and surpassed by only a few large enterprises (the Swatch Group, Richemont, Rolex). Controlling the new global value chains, from the production of watches in Hong Kong and China up to their distribution and sale throughout the world, has enabled the company built up by the Kartsotis brothers to carve out a name for itself as a world watchmaking giant, specialized in the field of fashion watches.

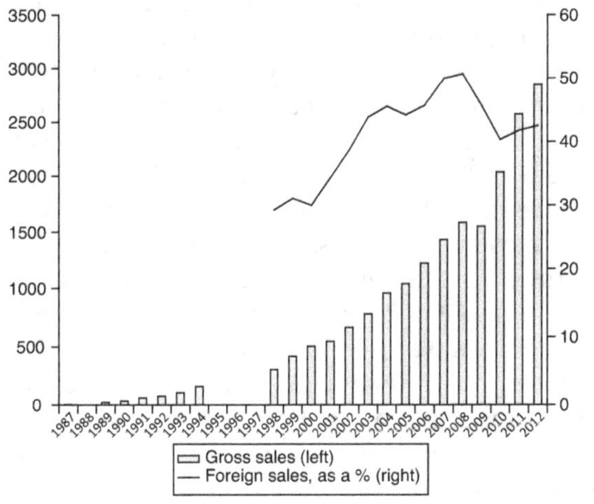

Figure 9.3 Turnover of the Fossil Group, in USD millions, and share of sales abroad as a percentage, 1987–2012

Note: Data unknown for 1988 and 1995–1997.

Source: Fossil annual reports and financial reports and the *International Directory of Company Histories*, Vol. 17, St. James Press, 1997.

However, the company does not provide details on its sales, and there is no information on watchmaking's share. Fossil's price segment and watch design make it a competitor for Japanese manufacturers and the Swatch Group brands like Tissot and cK Watch.

The movement war

This brief overview of the Swatch Group's main rivals has highlighted an issue which has become crucial: the manufacture of mechanical movements. Even though the Group's share of the volume of Swiss watch exports has remained stable (see Table 1.1, Chapter 1), it is growing in terms of value (72 per cent in 2010), and the Swatch Group has come to represent the main challenge as far as competition in the luxury watch field is concerned.

The transition to luxury has allowed the Swatch Group to transcend its historical role of a watch movement manufacturer and to concentrate on the design, production, and sale of watches in all market segments, primarily in the lucrative "affordable luxury" sector. First of all, the upmarket move of the entire Swiss watchmaking industry to the top end has led to a drop in the volume of exports, and thus to a stagnation in the amount of orders for movements or movement blanks manufactured by ETA SA. While the volume of watches exported by Switzerland averaged some 34.9 million pieces in the 1990s, it fell between 1998 and 2000, stabilizing at an annual average of 25.5 million pieces in 2001–2007. As far as the Swatch Group is concerned, this is a market contraction. Watch components (parts, movement blanks, and movements) sold outside the Group accounted for 17.1 per cent of turnover in 2000, 11.9 per cent in 2005, and only 7.6 per cent by 2010. Second, given the keener competition between watchmaking groups to control the market for finished watches, primarily in the luxury field, it became absurd for the Swatch Group to keep on supplying its main rivals.

Repositioning to the top end gave the Swatch Group an opportunity to free itself from its historical dependence on trade in watch movements. When it announced in 2002 that it would no longer supply parts and movement blanks to firms outside the Group after 2006, this triggered sharp reactions: it was estimated that the Swatch Group produced 80 per cent of all movement blanks and over half of all movements manufactured in Switzerland. In 2002, the factory

ETA SA had been investigated by the Swiss Federal Competition Commission (COMCO) for abuse of its dominant position and been forced to continue supplying its customers outside the Swatch Group. It hit back in December 2002 by announcing a steep price hike for the following year, particularly for high-end movements. This was precisely the segment in which the big Swiss groups were competing, and the Swatch Group had a clear advantage over brands which did not produce their own movements. Ultimately, COMCO and the Swatch Group reached an agreement in November 2006 whereby the latter undertook to keep selling its movement blanks to companies outside the Group until 2010. In that year, however, the Swatch Group's management announced its intention to continue implementing this policy and to gradually taper off its deliveries of movement blanks and watch parts from 2011 onwards, supplying only complete movements and assortments (watch regulating movements). Finally, in 2013, the Swatch Group announced its decision to phase out deliveries of assortments over time, a decision which COMCO opposed on the grounds that the Swatch Group's subsidiary Nivarox, which came out of ASUAG, had a virtual monopoly on this type of production.[8] For example, TAG Heuer's determination to limit its reliance on Nivarox for springs led the company to enter into an agreement with the Japanese watchmaker Seiko in 2012 to secure its supply of these parts.[9]

The Swatch Group's main competitors were no longer Japanese watchmakers; rather, they were other Swiss watchmaking firms, whose growth the Swatch Group had no reason to support. The calibres which ETA SA or Frédéric Piguet SA supplied to its rivals were encased as they were or slightly modified before being sold as luxury products all over the world. The ETA 2892 calibre, an automatic movement designed by Anton Bally and extremely widespread throughout the industry, is an excellent case in point (see Table 9.3). Sold at a price of around CHF100, it equipped the watches of many watchmaking firms competing with the Swatch Group.

ETA SA's plants clearly gave the Swatch Group a competitive advantage. However, this de facto virtual monopoly was primarily due to the fact that the other groups had failed to invest in the field of production. The patents for the majority of the most famous ETA SA movements, like the ETA 2892 automatic calibre or the Valjoux 7750 chronograph calibre, had fallen into the public domain, which meant

Table 9.3 Examples of the use of the ETA 2892 calibre

Company	Group	In-house reference / model
Breitling	independent	B45
Cartier	Richemont	Cal x49
Chronoswiss	independent	C. 9xx
Franck Müller	Franck Müller Watchland SA	GDT4600
Girard-Perregaux	PPR	GP 220–2200, 2280
IWC	Richemont	30110, 30710, 33110
Longines	Swatch Group	L619
Maurice Lacroix	Desco de Schulthess	ML05, 15, 51, 58, 91, 102, 103
Omega	Swatch Group	1120
Oris	independent	665
Piaget	Richemont	8532P
Tag Heuer	LVMH	Cal. 7
Ulysse Nardin	independent	UN 13, 20, 22, 26, 60, 97

Source: Brands' web pages, watchmakers' forum (http://forumamontres.forumactif.com, last accessed on 26 February 2011).

Table 9.4 Swiss mechanical movement makers, 2010

Name	Owner	Number of employees
ETA	Swatch Group	NA
Selitta	Miguel Garcia	200
Vaucher Manufacture	Parmigiani (75%), Hermès (25%)	300
SFT/Soprod	Festina	200
La Joux Perret	F. Wenger, E. Loth, R. Bohli	150
Dubois Depraz	Family Dubois	150
Technotime	Chung Nam (Hong Kong)	90
Christophe Claret	Christophe Claret	115
Concepto Watch Factory	Valérien Jacquet	100

Source: *Watch Industry*, Vontobel Equity Research, 2011.

that rivals could have invested in their production. However, apart from a few *manufactures*, such as Rolex, Audemars Piguet, and Jaeger-LeCoultre, only a handful of companies opted for this approach (see Table 9.4). The main examples were Selitta and Soprod, two former finishers for ETA SA who went into business on their own, offering movements based on ETA calibres. However, they remained heavily dependent on their former parent company, and these firms were two of the enterprises most strongly opposed to the Swatch Group's decision to stop delivering assortments and movement blanks over time.

10
Conclusion

Over a period of 15 years, between 1985 and 2000, the Swatch Group underwent profound change, which transformed it from a disparate conglomerate of weakly integrated Swiss watchmaking firms into a centralized, rationalized, and globalized multinational company. Whereas during the 1990s and the 2000s its Japanese rivals continued their stubborn race for technological innovation, which they saw as the driving force of a possible new phase of growth, the Swatch Group managed to establish itself as the world's leading watchmaking group without introducing any fundamental innovation in the technical field. This "non-technological innovation", as Jeannerat and Crevoisier call it,[1] was primarily based on two complementary policies: the rationalization of the production system, on the one hand, and the adoption of a new marketing strategy, on the other.

The production system was rationalized in an initial phase, which lasted from the Group's founding up to the late 1990s. Its main characteristics were the centralization of the production of movements and parts within the company ETA SA, the vertical integration of parts makers, and above all the globalization of supply, with the opening of production subsidiaries in Asia (Thailand, Malaysia, and China). Rationalization boosted the company's productivity and reinforced its historical role as a manufacturer and distributor of movements for the entire Swiss and world watch industries. Moreover, the sweeping changes introduced by the Swatch Group had a profound impact on the structure of this industrial sector, characterized by the transition from an organizational structure in the form of an industrial district to industrial concentration within a few multinational companies, the foremost of which was the Swatch Group.

As this rationalization of the production system was being finalized, a new marketing strategy emerged in the first half of the 1990s, in the wake of the Group's acquisition of Blancpain in 1992. Marked by a complete repositioning of its brands, this strategy was designed to turn th Swatch Group into a manufacturer, and above all a dealer, in complete watches. It was primarily characterized by two developments. First, the Group acquired several exclusive luxury brands in 1999–2000 (Breguet, Glashütte Original, Hatot, Jaquet Droz). Second, it chose Omega as its main brand to enter the global affordable luxury market, a segment which had traditionally been extremely profitable, and to compete with Rolex and Cartier. However, the Swatch Group is not only a producer of luxury watches. On the contrary, it is present in all market segments (entry, mid-level, affordable luxury, and exclusive luxury), with specific products for each segment. It is in fact the only watchmaking company in the world with a universal vocation. Its marketing strategy is closely linked to production-related matters and has a decisive influence on the Swatch Group's entire organizational structure.

The return of family capitalism in the watch industry

One of the most striking features of the historical development of the Swatch Group during its some thirty years of existence is undoubtedly the comeback of family capitalism it embodies. First of all, the entire concept of corporate governance was rethought, leading to a greater concentration of power. Since 1990, the Extended Group Management Board has operated as a marketing platform, whose goal is the coordination of a strategy applied globally. A noteworthy consequence in this respect is the strengthening of the company's family nature, with the growing involvement of the Hayek family in day-to-day operational management since 2000. Thus, in going global, the Swatch Group has become a family firm. This shows that globalized multinationals are not all companies which depend on financial markets, but that family entrepreneurship is capable of handling the operational management of a world-class company, as shown recently by Stéphanie Ginalski.[2] Moreover, this is a characteristic which the Swatch Group shares with other luxury multinationals, particularly the LVMH group, whose main shareholder and CEO is Bernard Arnault. Like Nicolas Hayek, Arnault has managed to build a global company through mergers and the rationalization of a series of small family firms. What is more, since the mid-2000s, Arnault has gradually been giving his children a greater role in the operational management of LVMH.[3]

This triumphant return of family capitalism in the Swiss watch industry is the outcome of a management policy aimed at ensuring the Group's financial independence and putting a stop to reliance on banks and the world of finance. The extraordinary growth in turnover, which exploded from CHF2.6 billion in 1995 to 4.5 billion in 2005 and 6.5 billion in 2010, has been accompanied by steady profit growth. Net earnings after taxes and amortization came to CHF273 million in 1995, 735 million in 2005 and 1.1 billion in 2010. Measured in relative terms, this works out to an average of 11.2 per cent of turnover in 1995–1999, 13.3 per cent in 2000–2004 and 16.4 per cent in 2005–2009. Thus, the Swatch Group's repositioning at the top end has had a direct impact in terms of increased turnover and profit margins. Profits have been generously distributed: the dividend rate soared from 17 per cent in 1995 to 55 per cent in 2000, 103.5 per cent in 2005, and 222.6 per cent in 2010. However, reserves were also set aside. Their total value came to CHF2 billion in 1995, 4.5 billion in 2005 and 7 billion in 2010. On the other hand, the shareholders' equity ratio remained stable, averaging 72 per cent for 1995–2009.

The bulk of the transition from an investor-oriented company to a family business took place in the 1990s. When the bankers and Schmidheiny left the Board of Directors, most of them were not replaced, and the number of Board members fell from nine to six in 1998. The three new members, all of whom were appointed in 1995, were Nayla Hayek, the daughter of the Board Chairman (1995), Professor Klaus Schwab, director of the World Economic Forum (1995), who resigned from his position two years later owing to a difference of opinion with Nicolas Hayek, and Ernest Tanner, CEO and Chairman of Swiss luxury chocolate maker Lindt & Sprüngli and a member of the Board of Directors of Crédit Suisse Group. Subsequent appointees (Johan-Niklaus Schneider-Ammann in 1999, Claude Nicollier in 2005, and Jean-Pierre Roth in 2010) have had little influence on the conduct of business. In addition, no one has been appointed to replace Peter Gross, who retired in 2011 for reasons of age. The appointment of astrophysicist Claude Nicollier, who is not a shareholder, could even be viewed as a marketing coup: by posing in his astronaut suit in the Group's annual reports and on its webpage, he recalls Omega's participation in the moon adventure, thereby reinforcing the brand's technical and historical legitimacy.

The strategy of increased financial independence had two effects. First, it decreased Hayek's reliance on investors and enabled him to strengthen his presence and power and that of his close family members, through the steady increase in the amount of shares held. The share of votes controlled by Hayek rose from 28.5 per cent in 1998 to 36.4 per cent in 2005 and 40.8 per cent in 2010. Second, this strategy of strengthening family control over the company was accompanied by a policy for the reduction of share capital. After being increased from CHF300 million to 331 million in 1993 to finance investments, share capital declined sharply over the next few years, falling to CHF137 million in 2002, a trend which allowed for greater shareholding control. It was against this backdrop that Nicolas Hayek prepared actively for the arrival of the second generation in the Group's management structure in the 1990s and 2000s. His daughter Nayla was appointed to the Board of Directors in 1995, while his son Georges Nicolas, called Nick, who worked in the marketing department of the Swatch brand, became a member of the Extended Group Management Board in 1994, then a member of the Executive Group Management Board in 1995. Shortly thereafter, he went on to become CEO of the subsidiary Swatch SA (1998). In 2003, Nicolas G. Hayek resigned as CEO of the Swatch Group, offering the position to his son Nick while remaining Chairman of the Board of Directors. In addition, Hayek's grandson Marc A. Hayek was appointed Marketing Manager of the subsidiary Blancpain SA (2001), then CEO at the age of 31 in 2002, a promotion which was no doubt linked to Biver's departure from the Swatch Group in 2003. Marc Hayek was also appointed to the Extended Group Management Board (2002), then to the Executive Group Management Board (2005). Thus, when Nicolas G. Hayek passed away in 2010, there was no break in terms of governance, as Nayla became Chair of the Board of Directors and Nick continued to serve as Chief Executive Officer.

From industrial district to leading firm

The Swatch Group's transformation from a multinational company to the world's leading watchmaking firm played a key role in Switzerland's comeback on the global watch markets. This transformation also marked a fundamental change in terms of its industrial organization and international competitiveness. The competitiveness of the Swiss watch industry on world markets in the late 1980s no longer relied on the flexibility of the traditional industrial district, as had been the case since the mid-19th century,[4] but rather on the presence of a world-leading firm. This change can be explained by the profound transformation of the market at the end of the 20th century, with a major shift from a fragmented and regionalized world market to a largely integrated global market, a trend which was particularly pronounced in the fashion and luxury goods trades. Until the 1980s, numerous small firms were able to find outlets for their specialized products thanks to the extremely fragmented nature of world markets. However, the globalization of brands and products made the organizational capabilities of multinational enterprises the key to competitiveness, even if they continued to use their geographical location as a resource for technology and marketing.[5]

There are two ways to view the relationship between the Swatch Group and the watchmaking district. First, one can examine the benefits for a globalized firm like the Swatch Group of being part of such a district. Second, the nature of the benefits conferred by the presence of a leading firm on the other enterprises of the district must be discussed.

As far as the Swatch Group is concerned, being inside the watch industrial district enables it to take advantage of two kinds of resources. The first is access to traditional know-how related to mechanical watchmaking. This technological resource, based on relations with legally independent subcontractors and on the employment of qualified workers trained in the country's various watchmaking schools, is often highlighted by scholars as one of the main reasons for the successful comeback by the Swatch Group and the Swiss watch industry as a whole on world markets.[6] However, know-how is not naturally rooted in a territory, as the literature generally illustrates. It is maintained and supported by social institutions such as the "Swiss

Made" legislation, hence the need for continuous training-related investment in Switzerland.

Moreover, the importance of the strategy of merging subcontractors and parts suppliers which was implemented in the mid-1980s must be underscored. The overall number of Swatch Group production subsidiaries in Switzerland (excluding watch assembly-makers) shot up from 7 in 1985 to 15 in 2010.[7] Consequently, there has been a strong tendency to internalize technical skills, such that the Swatch Group now controls all of the operations required to produce complete watches. When viewed from this perspective, the traditional ties of interdependence between firms within the district are no longer all that evident.

The second kind of resources available to the Swatch Group thanks to its presence in Switzerland is marketing resources, which have become a key issue on world markets. The Group's brands' transition to luxury goods has been accompanied by a reinforcement of their image as traditional, locally rooted products. Even though mechanical watches account for only some 20 per cent of overall exports, the communications policy of the Swatch Group – as well as that of its competitors – is based on a narrative highlighting tradition, brand heritage, and know-how. Accordingly, there are very rational reasons to keep production and facilities in Switzerland, which can explain why the Swatch Group has pushed hard since 2007 for a tightening of the specifications for the use of the "Swiss Made" label.[8]

It is more difficult to find evidence for the benefits which the industrial district itself derives from the presence of a leading firm. From the 1930s onwards, ASUAG played a major role as a provider of parts and movement blanks, a role taken over by the Swatch Group in the 1980s and 1990s. This in turn guaranteed the flexible and specialized character of the Swiss watch industrial district, as many small independent assembly-makers were able to emerge and grow. However, the Swatch Group's shift to producing complete watches and the creation of other concentrated luxury groups (Richemont, LVMH) led the Group to rethink its policy in 2002 and to decide to gradually stop supplying rival firms with parts and movements. An investigation by the Federal Competition Commission (Comco) notwithstanding, the Swatch Group has continued with this policy. Yet the presence of a leading firm is still profitable in the field of marketing. The various watch companies belonging to the Swatch

Group, such as Omega, become hubs for training managers, who go on to pursue their careers in other companies, taking with them new experiences and practices.

Watches, a Swiss product

The repositioning of watchmaking in the luxury segment and the Swatch Group's success on world markets are based on a communications policy aimed at tying watches to a territory, as much as they are on the rationalization and globalization of production systems. On the global luxury market, watches are Swiss products, rooted in a tradition and history symbolized by the image of the *manufacture*, or small-scale, high-end watchmaking workshop, which the Swatch Group strives to maintain. Accordingly, the Group's website (www.swatchgroup.com) and its annual reports contain virtually no references to production sites in Asia. Likewise, the Group stopped publishing figures on the number of Group employees in that part of the world in 2004.

Globalization and "Swissness" have been at the heart of the Swatch Group's strategy since the 1990s, an ambivalent attitude balanced on the "Swiss Made" legislation. The requirement that no more than half of the components of a so-called "Swiss Made" movement must be produced in Switzerland has made it possible to relocate aspects of production abroad and to boost Swiss watches' ability to compete on price. However, the strong push upmarket which came in the 2000s has enabled Swiss watchmakers to throw off the yoke of competition based on production prices, as the value of Swiss watches has become essentially emotional in nature. This no doubt explains the Swatch Group's strategy of buying up parts manufacturers and closing certain Asian production units in 2005–2006, a move designed to encourage the production of parts on Swiss soil. This is why the Group has been lobbying for more stringent "Swiss Made" protection measures since 2007. Even now, after it has become a multinational company, the Swatch Group continues to depend on a specific territory, Switzerland. Keeping alive a manufacturing tradition and using it as a marketing resource represents currently the Swiss watch industry's main competitive advantage on world markets.

Notes

1 Introduction

1. On the Japanese watch industry, see Donzé, "The hybrid production system"; Donzé, "*Rattraper et dépasser la Suisse*", *Seiko tokei*; and Shintaku and Kuwada, "Reorganizing mature industry".
2. Hara, "Kokusai kyoso"; Nakamoto, "Seihin no mojuruka"; and Sakakibara and Matsumoto, *Togogata kigyo no jirenma*.
3. Own calculation based on *Nihon no tokei sangyo tokei*.
4. Fragomichelakis, *Culture technique*; Tushman, *Rebirth of the Swiss Watch Industry*; Landes, *Revolution in Time*; Tajeddini and Trueman, "The potential for innovativeness".
5. Pasquier, "Une industrie remodelée": 315.
6. Zalio, "Territoire et concurrence des entrepreneurs": 260.
7. Glasmeier, *Manufacturing Time*: 247.
8. Jeannerat and Crevoisier, "Non-technological innovation".
9. Kebir and Crevoisier, "Cultural resources and regional development": 1198.
10. Munz H., "La fabrication et les usages politiques".
11. Donzé and Nishimura, *Organizing Technology Flows*.
12. *Illustrated Professional Dictionary of Horology*, Federation of the Swiss Watch Industry, http://www.fhs.ch/berner/ (last access: 23 December 2013).
13. Allères, *L'empire du luxe*.
14. Silverstein and Fiske, *Trading Up*.
15. Hoffmann and Hoffmann, "The pier framework of luxury innovation": 73.
16. Sicard, *Luxe, mensonge et marketing*: 21.
17. Daumas and de Ferrière le Vayer, "Les métamorphoses du luxe vues d'Europe".
18. Degoutte, "Stratégie de marques dans la mode".
19. Colli and Merlo, "Family business".

2 The Watchmaking Crisis of 1975–1985

1. *Forbes*, 1 July 2010.
2. *The New York Times*, 29 June 2010.
3. Piotet, *Restructuration industrielle et corporatisme*.
4. Piotet, *Restructuration industrielle et corporatisme*: 70.
5. Landes, *Revolution in Time*: 364–395.
6. Landes, *Revolution in Time*: 367.
7. Landes, *Revolution in Time*: 373.

8. See for example Schaller, "Les crises horlogères"; Perret, "Un canton chahuté"; Pasquier, "Une industrie remodelée"; and Pasquier, *La "recherche et développement" en horlogerie.*
9. Aguillaume, "Les horlogers suisses"; and Perret, Tissot e.a., *Microtechniques et mutations horlogères.*
10. Landes, *Revolution in Time*; Landes, "Swatch!"; and Pasquier, *La "recherche et développement" en horlogerie.*
11. Schumpeter, *The Theory of Economic Development.*
12. *Seiko tokei*: 154.
13. Pasquier, *La "recherche et développement" en horlogerie*: 379–390.
14. On the organization of the watch cartel, see Pasquier, *La "recherche et développement" en horlogerie*, Boillat, "État et industrie"; and Donzé, "Un cartel contre les transferts de technologie".
15. Pasquier, *La "recherche et développement" en horlogerie*: 437.
16. *Le Journal de Genève*, 24 October 1987.
17. Scheurer, *Les crises de l'industrie horlogère.*
18. "Timex".
19. *Kokusai tokei tsushin*, 1970: 477.
20. *Kokusai tokei tsushin*, 1970: 477.
21. Blanc, *Suisse-Hong Kong*: 148.
22. SSIH, *Rapport annuel*, 1973.
23. Hirano, *Seikoshashi*: annexes 32–36.
24. *Nihon keizai shimbun*, 14 October 1971.
25. *Nenpyo de yomu Seiko-epuson*: 32.
26. *Seiko tokei*: 96.
27. *Seiko tokei*: 103.
28. *Seiko tokei*: 132–133.
29. *Hong Kong Trade Statistics. Imports*, 1960–1965.
30. *Nihon tokei nenkan*, 1975.
31. *Hong Kong Trade Statistics. Imports*, 1975.
32. *Hong Kong Trade Statistics. Imports*, 1980.
33. Bolli, *L'aspect horloger*; Dirlevanger, Guex and Pordenone, *La politique commerciale de la Suisse* (especially chapter 4 on the "watch war").
34. *Swiss annual foreign trade statistics*, 1950–1959.
35. *U.S. Imports: Commodity by Country*, 1965.
36. Oxtoby, "The Role of Political Factors".
37. Donzé, "Institutionalizing 'useful knowledge'".
38. *U.S. Imports: Commodity by Country*, 1974–1976.
39. Kinugawa, *Nihon no baburu.*

3 The Creation of the Swatch Group and the "Swatch Legend"

1. JETRO, *Suisu.*
2. JETRO, *Suisu.*
3. Donzé, *History of the Swiss Watch Industry*: 76–83.

4. Koller, *"De la lime à la machine"*.
5. *Société générale de l'horlogerie suisse SA*.
6. *Swiss Federal Gazette*, 1931: 213.
7. *Industrie horlogère européenne*.
8. Richon, *Omega Saga*.
9. *Swiss annual foreign trade statistics*, 1970.
10. Fallet, "Nicolas Hayek"; and Wegelin, *Mister Swatch*.
11. *Journal de Genève*, 16 June 1984.
12. ASUAG, SSIH, and Swatch Group, *Annual Report*, 1974–1989.
13. Carrera, *Swatchissimo*; and Landes, "Swatch!".
14. Garel and Mock, *La fabrique de l'innovation*.
15. *Swiss annual foreign trade statistics*, 1983–2009.
16. *Swiss annual foreign trade statistics*, 1980.
17. Wegelin, *Mister Swatch*: 75–77.
18. Kapferer and Bastien, *The Luxury Strategy*: 234.
19. Morrison and Bouquet, *Swatch*.

4 rationalization and Globalization of Production Systems (1985–1998)

1. Richon, *Omega Saga*: 452–454.
2. Pasquier, *La « recherche et développement » en horlogerie*: 48–58.
3. Wengelin, *Mister Swatch*: 24.
4. *L'Impartial*, 23 January 1987.
5. *L'Impartial*, 17 August 1989.
6. *World Trade Atlas*.
7. *Journal de Genève*, 25 September 1997.
8. *Watches: China Sourcing Report*: 42–43; and Abrami e.a., *Fiyta*.
9. *L'Impartial*, 12 September 1994.
10. *Business Times (Singapore)*, 22 September 1994.
11. Statistics provided by the Citizen group.
12. Donzé, *History of the Swiss Watch Industry*: 119–120.
13. *World Trade Atlas*.
14. Oxtoby, "The role of political factors".
15. *L'Impartial*, 30 January 1973.
16. www.eta.ch (last accessed on 8 August 2013).
17. *Journal de Genève*, 11 April 1995; and Swatch Group, *Annual Report*, 1995.
18. *Journal de Genève*, 11 April 1995.
19. http://www.ronda.ch/fr/company/history/ (last accessed on 25 January 2012).
20. Own calculations based on *Swiss annual foreign trade statistics*, 1986–1987 and Swatch Group, *Annual Report*, 1986–1987.
21. Data provided by the Federation of the Swiss Watch Industry.
22. Donzé, "The changing comparative advantages".
23. *Le Temps*, 4 June 1999.
24. *The Business Times Singapore*, 16 March 2010.

25. Prahalad and Hamel, "The core competence": 90.
26. Chandler, *Inventing the Electronic Century.*
27. Heuer-Léonidas, *Annual Reports*, 1970–1982.
28. Heuer-Léonidas, *Annual Report*, 1976: 6.
29. *L'Impartial*, 29 June 1985.
30. Chisato, *Kigyo meihensen yoran*.
31. *Kaisha nenkan*, 1990–2010.

5 A New Marketing Strategy (1985–1995)

1. Zou and Cavusgil, "The CMS".
2. Hanssens, *Which International Marketing for Luxury Goods?*
3. Richon, *Omega Saga*: 71.
4. Richon, *Omega Saga*: 290.
5. *Officiel hommes* 65 (1987): 28.
6. Richon, *Omega Saga*: 39.
7. *L'Impartial*, 9 July 1992.
8. *Le Nouveau Quotidien*, 13 October 1991.
9. Henry Bédat, *Une région, une passion*.
10. *Journal de Genève*, 7 November 1980.
11. Hanssens *Which International Marketing for Luxury Goods?*: 37.
12. McCracken, "Who Is the Celebrity Endorser?"; and Zorik and Courvoisier, *L'horlogerie et ses ambassadeurs*.
13. Kapferer and Bastien, *The Luxury Strategy*: 262.
14. Quoted by Kapferer and Bastien, *The Luxury Strategy*: 263.

6 The Major Move into Luxury (Since 1995)

1. Truong, McColl and Kitchen, "New luxury brand positioning".
2. Richon, *Omega Saga*: 288.
3. Wegelin, *Mister Swatch*: 107.
4. Viredaz, "*Léon Hatot*".
5. Cardinal, *Abraham-Louis Breguet.*
6. Vanden Berghe, "Pierre Jaquet-Droz".
7. Fallet, "Piaget".
8. *L'Impartial*, 29 May 1997.
9. *Le Temps*, 27 January 2006.
10. *The Swatch Group.*
11. Global Trade Atlas.
12. *La Revue des Montres*, 2 July 2008.
13. Hausmann, "Le Chenit".
14. *Census of firms in 2001 and 2008.*
15. Own calculations based on the Employers' Federation, *Census*.
16. *Le Temps*, 7 May 2004.
17. *The Development and Innovation of the Hong Kong Watch.*

18. *Watches: China Sourcing Report*: 6.
19. Donzé, *History of the Swiss Watch Industry*: 134–135.
20. This share is estimated on the basis of the volume of Swiss watch exports, the Swiss domestic market being estimated at 10 per cent of exports and the volume of watchcase imports.
21. *Global Trade Atlas*.
22. http://www.ronda.ch/fr/company/history/ (last accessed on 25 January 2012).

7 Omega's Choice

1. Pasquier, *La « recherche et développement » en horlogerie*: 48–49.
2. SSIH, *Annual Reports*, 1965–1974.
3. Richon, *Omega Saga*: 47.
4. Richon, *Omega Saga*: 75.
5. Information provided by Aldo Magada, 5 April 2011.
6. Following biographical information taken from Richon, *Omega Saga*, and the social network www.linkedin.com (last accessed on 8 August 2013).
7. Kapferer and Bastien, *The Luxury Strategy*: 57.
8. Kapferer and Bastien, *The Luxury Strategy*: 94.
9. *Omegamania*.
10. http://www.omegawatches.com/spirit/history/extract-of-the-archives (last accessed on 10 August 2013).
11. *The Swatch Group*, Geneva: Helvea, 2007.
12. Hayek, *Au-delà de la saga Swatch*: 148.

8 China: A New El Dorado

1. Swiss foreign trade statistics, 2010.
2. Hanssens, *Which International Marketing for Luxury Goods?*
3. Swiss annual foreign trade statistics, 1930–1939.
4. *L'Impartial*, 8 November 1962, 28 May 1969, and 17 March 1975.
5. "Chugoku ni okeru tokei kogyo": 56.
6. Hanssens, *L'industrie horlogère est-elle assez chinoise?*: 12.
7. Business report of Guoco Capital Ltd, www.guococapital.com (last accessed on 20 January 2011).
8. *South China Morning Post*, 13 April 2006.
9. *Watch Industry*: 57.
10. Chevalier and Lu, *Luxury in China*: 146–147.
11. Chevalier and Lu, *Luxury in China*: 151–153.

9 Swatch Group's Competitors

1. www.businessweek.com (last accessed on 8 August 2013).
2. Donzé, *History of the Swiss Watch Industry*: 135–136.

3. Donzé, *History of the Swiss Watch Industry*: 138–139.
4. Hermès, *Rapports annuels*, 1994–2012.
5. LVMH, *Annual Reports*, 1999–2012.
6. Fossil Co., *Annual Reports*, 1998–2012; and "Fossil".
7. *L'Impartial*, 17 December 2001.
8. *Le Temps*, 13 July 2013.
9. *Le Temps*, 21 March 2012.

10 Conclusion

1. Jeannerat and Crevoisier, "Non-technological innovation".
2. Ginalski, "Can families resist managerial and financial revolution?".
3. Chatriot, "La construction récente des groups de luxe"; and Routier, *L'ange exterminateur.*
4. Linder, *De l'atelier à l'usine*, 2008.
5. Catalan and Ramon-Muñoz, "Marshall in Iberia".
6. See for example Pasquier, *La « recherche et développement » en horlogerie.*
7. Swatch Group, *Annual Reports*, 1985–2010.
8. *Bilan*, 24 April 2007.

References

Abrami, R. e.a. *Fiyta: The Case of a Chinese Watch Company*. Cambridge: Harvard Business School, 2008.

Aguillaume, C. "Les horlogers suisses face à la mondialisation (1968–1983)", *Cahiers de RECITS* (2004): 57–76.

Allères, D. *L'empire du luxe*. Paris: Belfond, 1992.

Bain & Company. *2012 Luxury Goods Worldwide Market Study*. Milan, 15 October 2012.

Bilan, 24 April 2007.

Blanc, J.-F. *Suisse-Hong Kong, le défi horloger. Innovation technologique et division internationale du travail*. Lausanne: Éd. d'En bas, 1988.

Boillat, J. "État et industrie: l'exemple du cartel horloger suisse (1931–1951)", in Cortat, A. (ed.) *Entreprises et cartels en Suisse. Etude de cas*. Neuchâtel: Alphil, 2010: 89–106.

Bolli, J.-J. *L'aspect horloger des relations commerciales américano-suisses de 1929 à 1950*. La Chaux-de-Fonds: La Suisse horlogère, 1956.

Business Times (Singapore), 22 September 1994 and 16 March 2010.

Cardinal, C. (ed.) *Abraham-Louis Breguet, 1747–1823: l'art de mesurer le temps*. La Chaux-de-Fonds: Institut l'homme et le temps, 1997.

Carrera, R. *Swatchissimo, 1981–1991: l'extraordinaire aventure Swatch*. Geneva: Antiquorum, 1991.

Catalan, J. and Ramon-Muñoz, R. "Marshall in Iberia. Industrial districts and leading firms in the creation of competitive advantage in fashion products", *Enterprise & Society* 14, no. 2 (2013): 327–359.

Census of Firms 2001 and 2008. Swiss Federal Statistical Office, www.bfs.admin.ch (last access 10 April 2012).

Chandler, A. D. Jr. *Inventing the Electronic Century: The Epic Story of the Consumer Electronics and Computer Industries*. Cambridge: Harvard University Press, 2001.

Chatriot, A. "La construction récente des groups de luxe français: mythes, discours et pratiques", *Entreprises et Histoire* 46 (2007): 143–156.

Chevalier, M. and Lu, P. *Luxury in China: Market Opportunities and Potential*. Singapore: John Wiley & Sons (Asia), 2010.

China Markets Yearbook. Beijing: Foreign Languages Press, 2005.

China Statistical Yearbook, Beijing: National Bureau of Statistics, 1980–1999.

Chisato, Y. *Kigyo meihensen yoran*, Tokyo: Nichigai Associates, 2006.

Chugoku shinshutsu kigyo ichiran. Tokyo: Sososha, 2013.

Compagnie financière Richemont SA, *Annual Reports*, 1999–2013.

Contrôle Officiel Suisse des Chronomètres (COSC), *Rapports annuels*, 1961–2010.

Convention patronale. *Recensement 2007*. La Chaux-de-Fonds: CP, 2008.

"Chugoku ni okeru tokei kogyo", *Nihon tokei gakkaishi* 118 (1986): 56–61.

Colli, A. and Merlo, E. "Family business and luxury business in Italy (1950–2000)", *Entreprises et Histoire* 46 (2007): 112–124.

Daumas, J.-C. and De Ferrière le Vayer, M. "Les métamorphoses du luxe vues d'Europe", *Entreprises et Histoire* 46 (2007): 6–16.

Degoutte, C. "Stratégie de marques dans la mode: convergence ou divergence des modèles de gestion nationaux dans l'industrie du luxe (1860–2003)?", *Entreprises et Histoire* 46 (2007): 125–142.

The Development and Innovation of the Hong Kong Watch and Clock Industry. Hong Kong: Hong Kong Watch Manufacturers Association, 2010.

Dirlevanger, D., Guex, S. and Pordenone, G.-F. *La politique commerciale de la Suisse de la Seconde Guerre mondiale à l'entrée au GATT (1945–1966).* Zurich: Chronos, 2004.

Donzé, P.-Y. "Un cartel contre les transferts de technologie: l'horlogerie suisse (1900–1970)", in Cortat, A. (ed.) *Entreprises et cartels en Suisse. Étude de cas.* Neuchâtel: Alphil, 2010: 137–163.

Donzé, P.-Y. "Institutionalizing 'useful knowledge': the Japanese External Trade Organization (JETRO) and the beginning of the Japanese Miracle – the case of the precision machine industry (1945–1960)", unpublished working paper given at the *Annual Conference of the European Business History Association*, University of Glasgow, 27 August 2010.

Donzé, P.-Y. *History of the Swiss Watch Industry from Jacques David to Nicolas Hayek.* Berne: Peter Lang, 2011.

Donzé, P.-Y. "The hybrid production system and the birth of the Japanese specialized industry: Watch production at Hattori & Co. (1900–1960)" *Enterprise & Society* 12, no. 2 (2011): 356–397.

Donzé, P.-Y. "The comeback of the Swiss watch industry on the world market: A business history of the Swatch Group (1983–2010)" *Journal of Strategic Management Education* 8, no. 2 (2012): 1–32.

Donzé, P.-Y. "Global competition and technological innovation: A new interpretation of the watch crisis, 1970s–1980s", in David, T., Mathieu, J., Schaufelbuehl, J. M. and Straumann, T. (eds) *Crises – Causes, interprétations et consequences*, Zurich: Chronos, 2012: 275–289.

Donzé, P.-Y. "The changing comparative advantages of the Hong Kong Watch Industry (1950–2010)", *Kyoto Economic Review* 81 (2014): forthcoming.

Donzé, P.-Y. *«Rattraper et dépasser la Suisse»: Histoire de l'industrie horlogère japonaise.* Neuchâtel: Alphil-Presses universitaires suisses, forthcoming in 2014.

Donzé, P.-Y. and Nishimura, S. *Organizing Technology Flows: Institutions, Actors, and Processes.* London & New York: Routledge, 2014.

Factory Directory in Thailand 2012/2013, Bangkok: Comm Bangkok Co., 2011.

Fallet, E. "Hayek Nicolas", *Dictionnaire historique de la Suisse*, www.dhs.ch (last accessed: 10 July 2012).

Fallet, E. "Piaget", *Dictionnaire historique de la Suisse*, www.dhs.ch (last accessed: 15 April 2011).

Feuille fédérale. Berne: Imp. Wyss, 1931.

Forbes, 1 July 2010.

"Fossil", *International Directory of Company Histories.* Detroit Mich. [etc.]: St James Press, vol. 17 (1997): 189–191.

Fragomichelakis, M. *Culture technique et développement régional: les savoir-faire dans l'Arc jurassien*. Neuchâtel: EDES, 1994.

Garel, G. and Mock, E. *La fabrique de l'innovation*. Paris: Dunod, 2012

Ginalski, S. "Can families resist managerial and financial revolutions? Swiss family firms in the twentieth century", *Business History* 55, no. 6 (2013): 981–1000.

Glasmeier, A. K. *Manufacturing Time: Global Competition in the Watch Industry, 1795–2000*. New York: The Guilford Press, 2000.

Global Trade Atlas, http://www.gtis.com (last accessed: 10 August 2013).

Hara, Y. "Kokusai kyoso to kodoka no inobeshon: waga kuni seizogyo no kyoso kiban", *Nagaoka daigaku kiyo* 2, (2003): 1–20.

Hanssens, N. *Which International Marketing for Luxury Goods? The Case of the Swiss Watch Industry in China*. University of Neuchâtel, unpublished master thesis, 2008.

Hanssens, N. *L'industrie horlogère est-elle assez chinoise? Analyse des stratégies d'utilisation d'ambassadeurs locaux et globaux*. University of Neuchâtel, Enterprise Institute, working paper, 2011.

Hausmann, G. "Le Chenit", *Dictionnaire historique de la Suisse*, www.dhs.ch (last accessed: 24 June 2013).

Hayek, N. G. *Au-delà de la saga Swatch: entretiens d'un authentique entrepreneur avec Friedemann Bartu*. Paris: Albin Michel, 2006.

Henry Bédat, J. *Une région, une passion: l'horlogerie: une entreprise: Longines*. Saint-Imier: Compagnie des Montres Longines SA, 1992.

Hermès, *Rapports annuels*, 1994–2012.

Heuer-Léonidas SA. *Rapports annuels*, 1970–1982.

Hirano, M. *Seikoshashi hanashi*. Tokyo: Seikosha, 1968.

Hoffmann, J. and Hoffmann, B. "The pier framework of luxury innovation", in Hoffmann, J. and Coste-Manière, I. (eds) *Luxury Strategy in Action*. London: Palgrave Macmillan, 2012: 57–73.

Hong Kong Trade Statistics. Imports. Hong Kong: Census and Statistics Department, 1960–1980.

L'Impartial, 8 November 1962, 28 May 1969, 30 January 1973, 17 March 1975, 29 June 1985, 23 January 1987, 17 August 1989, 9 July 1992, 12 September 1994, 29 May 1997, and 17 December 2001.

Industrie horlogère européenne: une expérience suisse: Ebauches SA – Lip, 1967–1973. Neuchâtel: Ebauches SA, 1973.

Jeannerat, H. and Crevoisier O. "Non-technological innovation and multi-local territorial knowledge dynamics in the Swiss watch industry", *Int. J. Innovation and Regional Development* 3, no. 1 (2011): 26–44.

JETRO. *Suisu no kinzokusei tokei bando shijo chosa*, Osaka: JETRO, 1982.

Le Journal de Genève, 4 November 1980, 16 June 1984, 24 October 1987, 11 April 1995, and 25 September 1997.

Kaisha nenkan. Tokyo: Nikkei, 1990–2010.

Kapferer, J.-N. and Bastien, V. *The Luxury Strategy: Break the Rules of Marketing to Build Luxury Brands*. London & Philadelphia: Kogan Page, 2012 (1st ed. 2009).

Kebir L. and Crevoisier O. "Cultural resources and regional development: The case of the cultural legacy of watchmaking", *European Planning Studies* 16, no. 9 (2008): 1189–1205.

Kikai tokei nenpo. Tokyo: MITI, 1960–2010.
Kikkawa, T. *Nihon sekiyu sangyo no kyoso ryoku kochiku*. Nagoya: Nagoya University Press, 2012.
Kikkawa, T. "International competitiveness of Japan's petroleum industry: A view from applied business history", *Kyoto Economic Review* 81, no. 1 (2012): 4–13.
Kinugawa, M. *Nihon no baburu*. Tokyo: Nihon keizai hyoron, 2002.
Kokusai tokei tsushin, 1970.
Koller, C. *« De la lime à la machine ». L'industrialisation et l'État au pays de l'horlogerie. Contribution à l'histoire économique et sociale d'une région suisse*. Courrendlin: CJE, 2003.
Landes, D. S. "Swatch! Ou l'horlogerie suisse dans le contexte mondial", in Bairoch, P. and Körner, M. (eds) *La Suisse dans l'économie mondiale*. Geneva: Droz, 1990: 227–236.
Landes, D. S. *Revolution in Time: Clocks and the Making of the Modern World*. Cambridge: Belknap Press, 2000 (1st ed. 1983).
Linder, P. *De l'atelier à l'usine: l'horlogerie à Saint-Imier. Histoire d'un district industriel: organisation et technologie. Un système en mutation*. Neuchâtel: Alphil, 2008.
LVMH, *Rapports annuels*, 2000–2012.
McCracken, G. "Who Is the Celebrity Endorser? Cultural Foundations of the Endorsement Process", *Journal of Consumer Research* 16, no. 3 (1989): 310–321.
Moët Hennessy Louis Vuitton (LVMH), *Rapports annuels*, 1999–2012.
Morrison, A. and Bouquet, C. *Swatch and the Global Watch Industry*. Cambridge: Harvard Business School, Case Study, 1999.
Munz, H. "La fabrication et les usages politiques du « patrimoine horloger » dans le Pays de Neuchâtel", www.ethnographiques.org (last accessed: 20 July 2013).
Nakamoto, D. "Seihin no mojuruka no shinten to gohatsu kigyo no keiei senryaku: seiko gurupu wo jirei toshite", *Yokohama kokusai shakai kagaku kenkyu* 8, no. 4 (2003): 147–164.
Nenpyo de yomu Seiko-epuson. Tokyo: Epson, 2001.
The New York Times, 29 June 2010.
Nihon gaikoku boeki tokei. Tokyo: Ministry of Finance, 1960–2010.
Nihon keizai shimbun, 14 October 1971.
Nihon no tokei sangyo tokei. Tokyo: Nihon tokei kyokai, 2010.
Nihon tokei nenkan. Tokyo: Somusho, 1975.
Le Nouveau Quotidien, 13 October 1991.
Omegamania: propriétés de divers amateurs qui seront vendues aux enchères au Mandarin Oriental Hotel du Rhône, samedi 14 et dimanche 15 avril 2007. Geneva: Antiquorum, 2007.
Oxtoby, F.-E. "The role of political factors in the Virgin Islands watch industry", *The Geographical Review* 60, no. 4 (1970): 463–474.
Pasquier, H. *La « recherche et développement » en horlogerie: acteurs, stratégies et choix technologiques dans l'arc jurassien suisse (1900–1970)*. Neuchâtel: Alphil, 2008.

Pasquier, H. "Une industrie remodelée (1975–2000)", in Tissot L. and Bujard J. (eds) *Le pays de Neuchâtel et son patrimoine horloger.* Neuchâtel: Edition de la Chatière, 2008: 307–315.

Perret, T., Tissot, L. e.a. *Microtechniques et mutations horlogères. Clairvoyance et ténacité dans l'Arc jurassien. Un siècle de recherche communautaire à Neuchâtel.* Hauterive: G. Attinger, 2000.

Perret, T. "Un canton chahuté", in Tissot L. and Bujard, J. (eds) *Le Pays de Neuchâtel et son patrimoine horloger,* Chézard-Saint-Martin: La Chatière, 2008: 301–305.

Piotet, G. *Restructuration industrielle et corporatisme: le cas de l'horlogerie en Suisse, 1974–1987.* Lausanne: Imprivite, 1988.

Prahalad, C. K. and Hamel, G. "The core competence of the corporation", *Harvard Business Review,* May–June 1990: 79–90.

La Revue des Montres, 2 July 2008.

Richon, M. *Omega Saga.* Bienne: Fondation Adrien Brandt en faveur du patrimoine Omega, 1998.

Routier, A. *L'ange exterminateur: la vraie vie de Bernard Arnault.* Paris: Albin Michel, 2003.

Sakakibara, K. and Matsumoto, Y. *Togogata kigyo no jirenma: Nihon tokei sangyo no seiko to satetsu.* New Energy and Industrial Technology Development Organization (NEDO), Discussion Paper Series 05–14, 2005.

Schaller, F. "Les crises horlogères, y a-t-il une spécificité?", in Cardinal, C. e.a. (eds) *L'homme et le temps en Suisse de 1291 à 1991.* La Chaux-de-Fonds: IHT, 1991: 273–278.

Scheurer, F. *Les crises de l'industrie horlogère dans le canton de Neuchâtel.* La Neuveville, 1914.

Schumpeter, J. *The Theory of Economic Development.* Boston: Harvard University Press, 1934 (first German edition 1912).

Seiko tokei no sengoshi. Tokyo: Seiko Co., 1996.

Shintaku, J. and Kuwada, K. "Reorganizing mature industry through technological innovation: De-maturing in watchmaking industry", in Hara T., Kambayashi N. and Matsushima N. (eds) *Industrial Innovation in Japan.* London & New York: Routledge, 2008: 101–117.

Sicard, M.-C. *Luxe, mensonge et marketing,* Paris: Pearson, 2010.

Silverstein, M. and Fiske, N. *Trading Up: Why Consumers Want New Luxury Goods and How Companies Create Them.* New York: Portfolio, 2005.

Société générale de l'horlogerie suisse SA. ASUAG. Historique publié à l'occasion de son vingt-cinquième anniversaire, 1931–1956. Bienne: Arts graphiques SA, 1956.

Société suisse pour l'industrie horlogère SA (SSIH), *Rapports annuels,* 1970–1982.

South China Morning Post, 13 April 2006.

Statistique annuelle du commerce extérieur de la Suisse, Berne: Direction générale des douanes, 1900–2012.

Swatch Group, *Rapports annuels,* 1983–2012.

The Swatch Group, Geneva: Helvea, 2007.

Tajeddini K. and Trueman M. "The potential for innovativeness: A tale of the Swiss watch industry", *Journal of Marketing Management* 24, no. 1/2 (2008): 169–184.

Le Temps, 4 June 1999, 7 May 2004, 27 January 2006, 21 March 2012, and 13 July 2013.

"Timex", *International Directory of Company Histories*. Detroit Mich. [etc.]: St. James Press, vol. 25, 1999, pp. 479–482.

Truong, Y., McColl, R. and Kitchen, P. J. "New luxury brand positioning and the emergence of *Masstige* brands", *Brand Management* 16, no. 5–6 (2009): 375–382.

Tushman, M. L. *Rebirth of the Swiss Watch Industry, 1980–1992*. Cambridge: Harvard Business School (case study 9–400–087/088), 2000.

U.S. Imports Commodity by Country, Washington, DC: U.S. Department of Commerce, Bureau of the Census, 1965–1976.

Vanden Berghe, M. "Pierre Jaquet-Droz, horloger mécanicien (1721–1790)", in Schlup, M. (ed.) *Biographies neuchâteloises*, vol. 1, Hauterive: G. Attinger, 1996: 154–158.

Viredaz, M. *"Léon Hatot et les horloges électriques ATO", Chronométrophilia 56, 2004*.

Watch Industry. Zurich: Vontobel, 2011.

Watches: China Sourcing Report. Hong Kong: Global Sources, 2005.

Wegelin, J. *Mister Swatch: Nicolas Hayek and the Secret of his Success*. London: Free Association Books, 2010.

Xinyu Hengdeli, *Annual reports*, 2004–2012.

Zalio, P.-P. "Territoire et concurrence des entrepreneurs: Marseille 1930", in Tissot L. e.a. (eds) *Histoire de territoires: les territoires industriels en question, XVIIIe-XXe siècles*. Neuchâtel: Alphil-Presses universitaires suisses, 2010: 253–304.

Zorik, K. and Courvoisier, F. (eds) *L'horlogerie et ses ambassadeurs*. Le Mont-sur-Lausanne: LEP, 2011.

Zou, S. and Cavusgil, S. T. "The GMS: A broad conceptualization of global marketing strategy and its effect on firm performance", *Journal of Marketing* 66 (2002): 40–56.

Index

Note: Page numbers referring to tables and illustrations are represented in bold.

www.ingramcontent.com/pod-product-compliance
Lightning Source LLC
Chambersburg PA
CBHW071604290626
47387CB00006B/1003
* 9 7 8 1 1 3 7 3 8 9 0 6 0 *